THE ITALIAN DIFFERENCE

TRANSMISSION

Transmission denotes the transfer of information, objects or forces from one place to another, from one person to another. Transmission implies urgency, even emergency: a line humming, an alarm sounding, a messenger bearing news. Through Transmission interventions are supported, and opinions overturned. Transmission republishes classic works in philosophy, as it publishes works that re-examine classical philosophical thought. Transmission is the name for what takes place.

THE ITALIAN DIFFERENCE
BETWEEN NIHILISM AND BIOPOLITICS

edited by
Lorenzo Chiesa and Alberto Toscano

re.press Melbourne 2009

re.press

PO Box 40, Prahran, 3181, Melbourne, Australia
http://www.re-press.org
© re.press 2009

This work is 'Open Access', published under a creative commons license which means that you are free to copy, distribute, display, and perform the work as long as you clearly attribute the work to the authors, that you do not use this work for any commercial gain in any form whatsoever and that you in no way alter, transform or build on the work outside of its use in normal academic scholarship without express permission of the author (or their executors) *and* the publisher of this volume. For any reuse or distribution, you must make clear to others the license terms of this work. For more information see the details of the creative commons licence at this website:
http://creativecommons.org/licenses/by-nc-nd/2.5/

British Library Cataloguing-in-Publication Data
A catalogue record for this book is available from the British Library

Library of Congress Cataloguing-in-Publication Data
A catalogue record for this book is available from the Library of Congress

National Library of Australia Cataloguing-in-Publication Data

Title:	The Italian difference : between nihilism and biopolitics / editors Lorenzo Chiesa, Alberto Toscano.
ISBN:	9780980544077 (pbk.)
ISBn:	9 780980666540 (ebook)
Notes:	Bibliography.
Subjects:	Political science--Italy--Philosophy. Nihilism. Biopolitics--Italy.
Other Authors/Contributors:	Chiesa, Lorenzo. Toscano, Alberto.
Dewey Number:	320.010937

Designed and Typeset by A&R

This book is produced sustainably using plantation timber, and printed in the destination market reducing wastage and excess transport.

CONTENTS

1 INTRODUCTION 1
Lorenzo Chiesa and Alberto Toscano

2 THE ITALIAN DIFFERENCE 13
Antonio Negri

3 FOUCAULT *DOCET* 25
Pier Aldo Rovatti

4 NIHILISM AS EMANCIPATION 31
Gianni Vattimo

5 COMMUNITY AND NIHILISM 37
Roberto Esposito

6 BEYOND NIHILISM: NOTES TOWARDS A CRITIQUE
OF LEFT-HEIDEGGERIANISM IN ITALIAN PHILOSOPHY
OF THE 1970S 55
Matteo Mandarini

7 THE SYMBOLIC INDEPENDENCE FROM POWER 81
Luisa Muraro

8 TOWARDS A CRITIQUE OF POLITICAL DEMOCRACY 97
Mario Tronti

9 CHRONICLES OF INSURRECTION: TRONTI, NEGRI
AND THE SUBJECT OF ANTAGONISM 109
Alberto Toscano

10 NATURAL-HISTORICAL DIAGRAMS: THE 'NEW GLOBAL'
MOVEMENT AND THE BIOLOGICAL INVARIANT 131
Paolo Virno

11 GIORGIO AGAMBEN'S FRANCISCAN ONTOLOGY 149
Lorenzo Chiesa

References 165

1 INTRODUCTION

Lorenzo Chiesa and Alberto Toscano

A decade and a half ago, Massimo D'Alema, then leader of the Partito Democratico della Sinistra, the successor to the Italian Communist Party (since dissolved, after a number of acronymic transformations, into the centrist Partito Democratico), wrote a 'manifesto' entitled *Un paese normale*, a normal country. In the present panorama of cultural, economic and political 'desolation',[1] that melancholy post-communist programme—Italy's secure entrance into the supposedly stable embrace of advanced European liberalism—is a dead letter. Indeed, it is in the mode of what we could call the biopolitically grotesque that Italy seems to make its occasional forays into the global media. The biometric census of Romani children, the formation of semi-legal vigilantes squads against phantasmatic foreign rapists, the legislative injunction that doctors report undocumented migrants to the police instead of curing them, a massive turnout for a bigoted and hypocritical celebration of national-Catholic hetero-normativity (called, in English, *Family Day*), the appointment of a former topless model to Minister for Equal Opportunities. The list could, and does, go on. Perhaps the most egregious instance of this mix of authoritarianism and bad taste was recently found in one of Berlusconi's press conferences apropos the case of Eluana Englaro—a woman who had spent seventeen years in a persistent vegetative state, and whose father had finally managed to legally se-

---

1. See Perry Anderson's recent acerbic two-part survey, 'An Entire Order Converted into What it Was Intended to End', *London Review of Books*, 31, 4, 26 February 2009, and 'An Invertebrate Left', *London Review of Books*, 31, 5, 12 March 2009.

cure the right to euthanasia only to be stopped by an 'exceptional' use of governmental power, with the connivance of the Catholic church. Commenting on the reasoning behind his 'pro-life' intervention, Berlusconi declared that Englaro was still a beautiful young woman, and that she could still bear children, as she was still menstruating. In passing, we could note that the theme of a biopolitics of exception, dear to Agamben, is here so grossly embodied that its Heideggerian *gravitas* seems to implode. The Italian exception in the epoch of the alleged generalization of the state of exception comes in the guise of a radical 'desacralization' of the very figure of *homo sacer*. Here, the spontaneous generation of mass-mediatic biopolitical pseudo-concepts, such as the recently coined '*fine-vita*' ('end-of-life'), appears to be inextricable from the vulgar spectacularization of the phenomena they attempt to describe.

But these indices of disaster, more obscene than obscure, are far from the ones that have often accompanied the perception of Italy, and of Italian thought, as somehow exceptional. The internal debate on the peculiarity or difference of Italy has a long history, being in many regards crucial to the theoretical contrasts within Italian Marxism. From abroad, leaving aside the pervasive clichés of a tourist political imaginary—easily impressed by, say, huge crowds of protesters at anti-government rallies—and the earlier awe at the singular cultural and political impact of the Italian Communist Party, this peculiarity has been captured, among others, in the images of a 'pilot-experience' and a 'laboratory'. The first, proposed by Guy Debord in his 1988 *Commentaries on the Society of the Spectacle*, paints the Italy of the 'years of lead' of political violence as the cutting-edge in new forms of manipulative repression, covert action and spectacular conformity. The second, outlined by Michael Hardt in an important collection of 'post-workerist' radical theory, finds, in the Italy of *operaismo* and *autonomia organizzata* 'a kind of laboratory for experimentation in new forms of political thinking', albeit one whose exceptionality comes to a close as Italy 'converges' with other countries through the 'postmodernization of the economic realm and the Americanization of social and cultural fields', i.e. through what Hardt and Negri would later dub 'Empire'.[2] But is

---

2. Michael Hardt, 'Introduction: Laboratory Italy', in Paolo Virno and Michael Hardt (eds.), *Radical Thought in Italy: A Potential Politics*, Minneapolis, Univer-

this motif, of the baleful or joyful 'advantages of backwardness' still available for use? Was it ever truly convincing? Isn't it simply the other side of what Italian historians and political scientists refer to as the *ritardo storico*, the historical delay of Italy as a modern liberal democracy, which itself supposedly allowed the recently united country to 'pilot-experience' Fascism? If a kind of Italian exceptionalism (rather than an unproblematic specificity) can't be called upon to sanction the idea of an 'Italian difference' in political thought, what can?

This collection takes its cue, and its title, from a pamphlet by Antonio Negri. In the dismissal of *pensiero debole* (weak thought) and in the retort by Pier Aldo Rovatti, one of the promoters of that current, we can identify the main polemical axis that structures this volume, as well as some of the stakes of thinking the political and theoretical 'difference' of the Italian situation.[3] Negri's plea for a muscular political ontology of revolutionary subjectivity and creative difference—which references Antonio Gramsci, Mario Tronti and Luisa Muraro as the three isolated pinnacles of Italian twentieth-century thought—evinces the short-circuits, so prevalent in much of the Italian debate, between the political, the metaphysical, the cultural, and even the personally anecdotal. The alleged contrast between a 'weak' and a 'strong' thought manifests some of the paradoxical features of the politico-philosophical debates on the Italian Left, above all the peculiar admixture of the extremely parochial (the combination of debates on first-name terms and idiosyncratic political trajectories) and the intensely universal (the attempt to address Politics, Being, Humanity). It also demonstrates the conflicted influence of other strands of European thought on the Italian scene. Whilst Negri has frequently alluded to washing the linen of Italian (workerist) Marxism in the (Foucauldian and Deleuzian) waters of the

sity of Minnesota Press, 1996, pp. 1 and 5.

3. It is worth noting that the two main extant English-language collections of Italian philosophy have the deconstructivist and post-Heideggerian area of *pensiero debole* as their principal point of reference, see Giovanna Borradori (ed.), *The New Italian Philosophy*, Evanston, Northwestern University Press, 1989, and Silvia Benso and Brian Schroeder (eds.), *Contemporary Italian Philosophy: Crossing the Borders of Ethics, Politics and Religion*, Albany, State University of New York, 2007. In Italian, the vicissitudes of a philosophical field putatively dominated by a post-Heideggerian questioning of nihilism is dealt with in Giuseppe Cantarano, *Immagini del nulla. La filosofia italiana contemporanea*, Milano, Bruno Mondadori, 1998.

Seine, Rovatti objects to Negri the fact that these French references (to which one should certainly add Derrida) are the very ones that accompanied the moment of *pensiero debole*. The significance of Heidegger on the mutations of Italian thought is also at stake: intriguingly, while Negri reproaches *pensiero debole* for its Heideggerianism (a theme explored in genealogical detail in Mandarini's contribution to this volume), this is something that determines Vattimo's position, but not Rovatti's. On the other hand, Agamben, who does not suffer the attack levied against *pensiero debole* is much more emblematic of a Heideggerianisation of the French 'post-structuralist' legacy, and in particular of Foucault's late 70s thematisation of biopolitics and governmentality, not to mention the fact that he himself aligns his philosophical position with a thought of 'weakness'.[4] This does not prevent Negri's understanding of the biopolitical field as a 'constituent affirmation' of creative difference from constructively dialoguing with the *Angelus Novus*' insistence on the thanatological destiny of modern Western politics.

But behind these struggles over political metaphysics we can also see the acrimony generated by different reactions to the counter-revolution,[5] ebb or transformation represented by the 1980s, the very period of Negri's Parisian exile and of the formulation of *pensiero debole*. In this regard, it is worth reflecting on the pertinence, whether despairing or deflationary and ironic, of the notion of nihilism to that moment in Italian thought (the period which not only saw the publication of the collection *Il pensiero debole*, but also of Agamben's *Il linguaggio e la morte*, Cacciari's *Icone della legge*, and Negri's books on Spinoza and Leopardi). As so often with these issues, despite the conceptual and philological rigour brought to bear by the likes of Esposito, it is very difficult indeed to disentangle the 'local colour' (the expe-

---

4. See Giorgio Agamben, *The Time That Remains: A Commentary on the Letter to the Romans*, Stanford, Stanford University Press, 2005, pp. 136–7.

5. See Paolo Virno, 'Do You Remember Counter-Revolution', in Virno and Hardt (eds.), *Radical Thought in Italy*. Virno's essay uses the axiom of the 'primacy of resistance', so crucial to workerism and post-workerism (see Toscano's contribution to this volume) to argue in detail for the fact that Berlusconi's rise can only be understood as a perverse hijacking of the collective tendencies that marked the 'Red Decade' of '68–'77. What we are to make of the perpetuation of this counter-revolution in the apparent absence of the antagonistic forces it perverted is not clear.

rience of Craxi's opportunistic socialism, the incarceration and self-destruction of the extra-parliamentary Left, the anthropological transformation—or even, following the prescience of the late Pasolini, the 'anthropological genocide'—effected by Berlusconi's TV culture...) from philosophical writings steeped in an erudite and inquisitive philosophical culture, where a long-term allegiance to German philosophy was complemented or perverted by the new waves of thought from France. Ontological and political nihilism, like ontological and political affirmationism, often seem indiscernible. Consider also the increasing significance of Christian and Catholic thematics—in the guise of Agamben's and Negri's divergent Franciscanisms (see Chiesa's contribution to this volume), Vattimo's full assumption of his *catto-comunismo* (Catholic-Communism), or Muraro's feminist fondness for Benedict XVI's views on sexual difference, though we could also add, in a more attenuated guise, Virno's reliance on the theme of revelation, Tronti's arguments on prophecy and the *katechon*, or Esposito's reflections on birth. Whether we are considering biopolitics, nihilism or the vicissitudes of post-Christian subjectivity, recent radical Italian thought confronts us with a parallax view or disjunctive synthesis of national and conjunctural idiosyncrasies, on the one hand, and a series of potent theoretical abstractions that have a remarkable capacity for 'travelling', on the other. At the level of its international impact, the combination of a strong tendency to epochal periodisation (as applied to the notions of biopolitics, nihilism or Empire) and a proliferation of meta-political subjects or figures (*Muselmann*, refugee, multitude, exodus, up to the tourist),[6] mainly forged in a period of political retreat or defeat, have allowed the theoretical 'laboratory Italy' a remarkable capacity to speak—frequently through the medium of radical misunderstanding—to a bafflingly disparate set of situations. It is all too easy to imagine a Reading Agamben in Bogotà, a Reading Negri in Tehran, a Reading Vattimo in Beijing, a Reading Esposito in Seoul .... Though such a sociology of philosophy is beyond our remit, it would be worth considering the difference between this phenomenon of diffusion and that of

---

6. In his recent *Il Regno e la Gloria: Per una genealogia teologica dell'economia e del governo* (Milano, Neri Pozza, 2007, p. 158), Agamben speaks of the tourist as 'a figure whose "political" meaning is consubstantial with the prevailing governmental paradigm' on the basis of his 'irreducible extraneousness with regard to the world'.

French, or German philosophy. At the national level, it is an open question whether theoretical interventions—such as, in this volume, Muraro's feminist interrogation of power, Virno's sensitivity to the ambivalence of political anthropology, or Tronti's sober estimation of Italy's democratic *embourgeoisement*—can serve to counter the 'desert' evoked by both Negri and Esposito in their essays, the desolation with which we began—a conjuncture that brings the themes of nihilism, biopolitics and Christianity into a particular appalling configuration.

This volume brings together a number of texts by different generations of Italian thinkers which address, whether in assertive, problematising or genealogical registers, the entanglement of philosophical speculation and political proposition within recent Italian thought. It is not by any means comprehensive, nor does it define and determine a specific debate, but it will hopefully allow the reader to discern a constellation of themes and problems—biopolitics, nihilism, militant subjectivity, political anthropology—which, whilst stamped by their origins in a determinate political situation, continue, through the power of their abstractions, to influence an international theoretical debate (albeit one which, it must be noted, is itself marked by its Atlantic mediations; the global success of books such as Negri's *Empire* is often, correctly or incorrectly, perceived in Italy as yet another by-product of US cultural hegemony).

We begin with Antonio Negri's pamphlet, 'The Italian Difference', which casts a polemical eye on the panorama of twentieth-century Italian philosophical culture and declares that only three figures stand as exceptions to a pervasive political and intellectual capitulation: Antonio Gramsci, Mario Tronti and Luisa Muraro. Negri argues that the two key post-war contributions to an Italian political ontology, the workerism of Tronti and the feminism of Muraro, start from the identification of the principal forms of exploitation, capitalism and patriarchy, to develop a potent thinking of singularity and creative difference. He concludes that they provide the basis for a political philosophy of the multitude that can at last move beyond postmodernity.

In a wry response to Negri's article, Pier Aldo Rovatti—one of the key figures behind the *pensiero debole* movement attacked by Negri in 'The Italian Difference'—defends the Foucauldian inspiration behind his own understanding of philosophy. He points to the anachronism of the national image of thought put forward by Negri in his article and questions his interpretation of the problem of difference. Rovatti disputes the idea that philosophy can synthesize by *fiat* different expressions of subjectivity into a unitary political subject, and calls for a reflexive clarification of the tasks of the philosopher, one that would not end up recreating a logic of mastery.

Is the philosophical idea of nihilism compatible with a project of emancipation based on concepts such as autonomy, equality and freedom? This is the question to which Vattimo's contribution seeks to provide a response. For Vattimo, the notion of nihilism is inseparable from that of hermeneutics, understood as the historically situated character of universal claims. Rather than undermining emancipation, for Vattimo, a nihilistic hermeneutics is precisely what frees us from foundations, and should thus be understood as an emancipatory force. The article tries to counter a purely tragic understanding of nihilism with the constructive political horizons opened up by a nihilistic hermeneutics, which allows us to think anew the ideas of freedom and equality.

Developing the arguments put forward in books such as *Communitas*, in 'Community and Nihilism' the political philosopher Roberto Esposito tries to overcome the customary opposition between the notions of community and nihilism. His aim is to rethink what community might mean in an age of 'completed nihilism'. In a subtle genealogical and etymological analysis of the concept of community, he demonstrates how, rather than establishing a substantial and positive bond, community is constituted by nothingness, by a shared lack—which communal, communitarian and totalitarian politics seek to deny. The excavation of the meaning of *communitas* allows Esposito to critically examine the manner in which the thinking of community has been expunged by modern political philosophy.

Matteo Mandarini's article, 'Beyond Nihilism: Notes Towards a Critique of Left-Heideggerianism in Italian Philosophy of the 1970s', provides a much-needed introduction to the philosophical debates around nihilism and negative thought which preoccupied many Italian Left intellectuals in the seventies, and which still have important repercussions today. In order to present the principal stakes of the 'Left Heideggerian' current, the article undertakes a close reading of Massimo Cacciari's 1976 book *Krisis*, and of Antonio Negri's critical response to it—first in a review of the book, and then in a number of texts from the seventies and eighties, closely analysed by Mandarini, in which Negri develops a positive political metaphysics. This contrast between Cacciari and Negri allows Mandarini to investigate the significance of seemingly recondite philosophical issues to the development of Italian radical political thought, and to identify some of the key stakes of this debate: the status of politics and the political, the role of ontology, the place of dialectics and, crucially, the opposition between Cacciari's formalistic understanding of negativity and Negri's link between negativity and antagonism.

In her essay 'The Symbolic Independence from Power', Luisa Muraro begins from the philosophical question of the 'unthought', and asks how our very image of thought is transformed when the thinking subject is a woman, and her thought is specifically linked to the experience of a body. On the basis of a feminist interrogation of sexual difference which reveals the forms of violence inherent in certain claims to universality, Muraro tries to develop a thinking of politics which would rest on its symbolic distance or independence from power. Through readings of Freud, *Macbeth*, Saint Paul and women's narratives, Muraro investigates the dangers borne by the fusion of power and politics and explores the ways in which they could be disjoined.

Starting from the idea that democracy always binds together practice of domination and project of liberation, Tronti formulates the conditions for a critique of democracy that would permit a rebirth of political thought in the current conjuncture. Bringing the heterodox Marxist traditions of 'workerism' and the 'autonomy of the political' together with the feminist thinking of difference, Tronti underscores the identitarian tendencies of democracy and the difficulties of combining democracy with a genuine notion of freedom. For Tronti, democracy is increasingly synonymous with the pervasiveness of capitalism understood as 'bourgeois society', and the victory of 'real democracy' (as one might speak of 'real socialism') is the sociological victory of the bourgeoisie. The *homo oeconomicus* and the *homo democraticus* are fused into the dominant figure of democracy, the 'mass bourgeois'. Against the depoliticizing consequences of 'democratic

Empire', Tronti proposes a profound rethinking of our notion of politics, one which should not shy from reconsidering the elitist critiques of democracy.

Alberto Toscano's contribution seeks to trace the origins of contemporary 'post-workerism' in the formulation of concepts of political subjectivity, antagonism and insurrection in Tronti and Negri. In particular, it tries to excavate the seemingly paradoxical position which postulates the increasing immanence of struggles, as based on the Marxian thesis of real subsumption, together with the intensification of the political autonomy or separation of the working class. In order to grasp the political and theoretical proposals of Italian workerism and autonomism, Toscano concentrates on the thesis of a historical transformation of capitalism into an increasingly parasitical and politically violent social relation, a thesis which is grounded in an interpretation of Marx's notion of 'tendency' and which serves as the background to the exploration, especially in Negri, of increasingly uncompromising forms of antagonism. The article focuses especially on Tronti's so-called 'Copernican revolution'—giving workers' struggles primacy in the understanding of capitalism—and critically inquires into the effect of this workerist axiom on Negri's writings on proletarian sabotage and insurrection in the 1970s. By way of a conclusion, it notes the difficulties in prolonging the workerist gambit in light of capital's continued effort, as Tronti would put it, to emancipate itself from the working class.

In 'Natural-Historical Diagrams: The "New Global" Movement and the Biological Invariant', Paolo Virno puts forward the thesis that the contemporary global movement against capitalism, and the post-Fordist regime it is responding to, is best understood in terms of the emergence of 'human nature' as the crux of political struggle. According to Virno, the biological invariant has become the raw material of social praxis because the capitalist relation of production mobilizes to its advantage, in a historically unprecedented way, the species-specific prerogatives of *homo sapiens*. Through the concept of 'natural-historical diagrams', the article explores the significance of socio-political states of affairs which directly display key aspects of anthropogenesis, and, making use of Ernesto De Martino's concept of 'cultural apocalypses', considers the different relations that a biological 'background' and a socio-political 'foreground' entertain in traditional and contemporary societies. The attempt to develop a 'natural history' of such diagrams leads Virno to reflecting on the importance of the language faculty, neoteny, non-specialization and the absence of a predetermined natural environment for political action. This reflection on the contemporary importance of political anthropology leads Virno to a set of concluding remarks on the role of ethics and the idea of the 'good life' in the practice of the 'new global' movement.

The final paper, by Lorenzo Chiesa, analyses Agamben's notion of *homo sacer*, showing how it should not be confined to the field of a negative critique of biopolitics. In his work, Agamben cautiously delineates a positive figure of *homo sacer*, whom, according to him, we all virtually are. Such a figure would be able to subvert the *form* in which the relation between bare life and political existence has so far been both thought and lived in the West. How and when is this passage from negative to positive sacredness historically accomplished for Agamben? Is such transit after all thinkable? These are the two basic questions he both unintentionally formulates and leaves undecided in his book *Homo Sacer*. Agamben further elaborates his investigation of biopolitics in the book he dedicates to Saint Paul, *The Time That Remains*. Chiesa suggests that, in the latter volume, the figure of *homo sacer* as earthly hero is tacitly transposed onto that of the messianic man. This can only be achieved by means of a detailed Christian—and more specifically Franciscan—development of the ontological notion of 'form of life'. Problematically enough, Agamben is able to carry out a transvaluation of biopolitics only in the guise of a bio-*theo*-politics.

2 THE ITALIAN DIFFERENCE

Antonio Negri

When one says 'philosophy', one means that critical activity which allows one to grasp one's time and orientate oneself within it, creating a common destiny and witnessing the world for this purpose. If one defines it in this way, after Giovanni Gentile and perhaps a bit Benedetto Croce, there hasn't been any philosophy in Italy in the twentieth century. With a couple of important exceptions (three, to be more precise).

Before considering the exceptions, let us however look at the development of Italian philosophical thought as it was outlined in the twentieth century, as such and in the European context. The Italian nineteenth century was endowed with great philosophical personalities: Leopardi, Rosmini, De Sanctis, Labriola... However, these were personalities and almost never schools, because 'Italy did not have a centre', because, given its historico-political situation, communication was fragmented, which prevented the formation of a public space. Just as Hegel used to say that 'Germany doesn't have a metaphysics, it doesn't have a temple', so Leopardi's bitter statement denies the existence of an Italian public space—not a hegemonic centre, but simply the public nature of communication. As a consequence, that nineteenth-century flash of philosophical activity did not find any continuity in the twentieth century. Philosophy did not go beyond the Risorgimento. The ballast of universities, the pandemonium of fashions, the frivolity of the new media tools: all this asserted itself in the passage of the century, creating and spreading dogmatic philosophical visions, sectarian gratifications or literary digressions. However, if Italy does not have a centre, Italian philosophy is not even provincial: it is just weak [*debole*], it has always been a weak philosophy, weak in the face of politics and bosses, dictators and popes.

In the twentieth-century decline of ideas and debates, the vilest point was perhaps reached when some, with a certain pride, proclaimed their thought and their definition of contemporary philosophy as 'weak'. Others named it more properly 'limp thought' [*pensiero molle*]. It looked like an attempt to acclimatize the postmodern beneath the lukewarm Italic sky: actually, it was a plan to flatten the richness of the articulations and surfaces of the real, the *dispositifs* and *agencements* of French poststructuralist critique, onto the horizon of Heideggerian ontology. More deceitfully, it was a plan to repudiate the history of the insurgences and resistances that had accompanied the first construction from below of a public space in Italy, the first democratic construction after Fascism. After 1968, the power[1] of the struggles and of the new massification of political discourse needed to be delicately confined in a (far from delicate) renewed ontology of fascism. Weak thought translated into Italian a Foucault and a Deleuze dressed like game-show hostesses; it made them dance on the cultural pages of so-called 'Leftist' newspapers, especially *La Repubblica*... We received a special treatment: the new composition of labour, in its immaterial and intellectual figure, was presented to us as evasive, aleatory. Its creativity was mystified in the illusory figure of an 'end of history', which was supposed to mean the disappearance of wage-labour and the working class. The tragedy that accompanied the mutation of the industrial plan and the metamorphosis of labour-power was thus led back and closed into the inconsistent mess of a supposed defeat of communism and a presumed triumph of the '*Milano da bere*'.[2] Limp thought and the Craxi era go together: however, one should admit that Craxi's Proudhonism was by all means weightier and philosophically more relevant than Vattimo's and Ferrara's light thought.

From the Right, the ineffable cultural pages of conservator-

---

1. Unless otherwise specified, 'power' always translates '*potenza*'. [*Translator's note*]

2. In the early 1980s, '*Milano da bere*'—literally 'Milan to drink'—was the slogan of an advertising campaign for a popular liquor. Such slogan came to epitomize the mundane dimension of Milan as a city of fashion, media, and glamour. [*Translator's note*]

ism considered even limp thought too *risqué*: there the Mitteleuropean necrosophy of the Claudio Magris Co. ruled and continues to do so. For a long time, looking through *Il Corriere della Sera*'s cultural pages was like observing the malaise of a club of spinsters from Lower Saxony or, even worse, the unhappiness of a small community of Romanian Jews. In contrast to the uncontaminated flow and lightness of limp thought, the Danube presented itself as a viscid and heavy river. Too bad for the Danube, which really doesn't deserve this! Too bad, really: mourning that intelligence which did not manage to resist Nazism, could not in fact cancel (as the ambiguous mentors of Mitteleuropa so ardently wished) the force of the historical process, collective and not individual, communist and not liberal, which defeated it! Mitteleuropa is also resistance.

This was the shape of Italian philosophy between Right and Left until the Nineties, and even until the present day. Occasionally, after 1989, those cultural pages were flooded by the wave of reactionary apologists, of those historians (who do not have anything to do with Droysen or Braudel, despite boasting to be their heirs) who claimed that revisionism had a right to tell how things had really happened. Occasionally, the cultural pages were also criss-crossed by vague neo-transcedentalist tendencies. Habermas and Rawls were welcomed into our intellectual culture, since they showed that one could be a radical when young, but necessarily became a reactionary when old....

So why has Italian philosophical culture—together with the cultural pages that express it and the intellectuality that basks in it—duplicated the glitter of a Raffaella Carrà's variety show?[3] There's something wrong in this story, especially when one considers that this centre (that of limp philosophy, of the televisual and journalistic degeneration of cultural communication) has been the only centre that the *Bel Paese* has had since Fascism.

Stop. Let's move on to the exceptions: three, as we've said. The first one was Gramsci: the hunchback, the betrayer of Stalinism, the one whom the other political prisoners used to pelt with stones in jail. Gramsci re-established philosophy where it should have remained, in the life and struggles of ordinary people. He

---

3. Raffaella Carrà is a popular Italian TV hostess. [*Translator's note*]

reinvented Gentile, attempting to turn actualism into the basis of a thinking and praxis of the future (in a rather far-fetched way, one has to admit). This was not an exhilarating adventure: a man of the Left, a communist who puts Gentile's philosophy back on track, partly remains a man of the nineteenth century... Gramsci was just such a man, and therefore represents the true continuity of the Risorgimento in twentieth-century Italy. Unfortunately, the non-philosophy of Togliatti's epigones (which is to say, the horrible cynicism that today has become hegemonic in the Left) and the exterminating voluptuousness of Stalinists (which expressed itself so well against movements in the 1970's and which is still out there, as frenzied as ever) have hidden and mystified even this poor revolutionary voice. Sorry, not just revolutionary, but lively, intelligent, generous; in the philosophy of life that opposes the ontology of death, there is always a certain creative aspect. This is precisely what they (the bosses, the power *élites*) do not want. The Gramscian exception has thus been reduced to an experience rooted in the past and perhaps only able to prefigure a utopian future: on the contrary, it was a break, it was resistance.

From the outset we've said that there are two other exceptions, two other fundamental breaks, not only in the continuity of the history of Italian philosophy between the nineteenth and twentieth century, but, at the same time, in the material texture of the intellectual life of Italian society, in its politico-linguistic structure. What are these two other exceptions? Do they allow us to say that this long period of time which prepared us for the year 2000 had a constructive, creative, and innovative aspect; that it represented a power on which we can rely? Let's attempt to answer these questions.

The first exception that the Italian twentieth century witnessed, the first philosophical and political force able to plunge its hands into the real and again grab hold of the Risorgimento and the anti-capitalist powers of the origins—well, this exception was workerism, the work of Mario Tronti. In addition, there was another exception contemporary to workerism, almost hidden and yet which operated profoundly: this is the feminist thought of difference developed by Luisa Muraro. These are the two only elements of theoretical innovation in twentieth-century Italic ontology [*ontologia italica*]. Both move from the consideration of the fundamental forms of the constitution of exploitation, of man over man and of man over woman. Thus, there are two forms of exploitation: capitalist and patriarchal. Philosophical thought can only be born—and sustains itself in both these cases—when it focuses on the biopolitical theme of reproduction. We are thus at the centre of the postmodern figure of philosophical reflection. While Aristotelian being descends from the whole to individualities and then re-ascends from individualities to the One (by means of the modes of causation), the postmodern does not accept the 'upward path' or 'downward path' as genealogical and productive; it does not even accept individuality, but only the singular: it therefore considers *difference* as the creative *quid* that stretches between nature and history. Workerism and the feminism of difference were born in the 1960s from the opportunity opened by the enormous development of struggles; in these struggles, irreducible differences were posed, as new subjectivities were formed both in the workers' battle against waged labour and in the feminine insurrection against patriarchal domination. It was the discovery of these differences that determined the rebirth of philosophy. It is resistance that produces philosophy.

Having defined their shared birthplace, let's look at what these philosophies have in common. In the first place, these two positions fight against dialectics. There is no longer any possible recomposition or *Aufhebung*... 'Let's spit on Hegel', Carla Lonzi used to say. Let's break any narrative connection that doesn't know immediately how to develop class struggle, Alberto Asor Rosa added. While dialectical arrogance claims to lead everything back to the One, here everything is instead fixed upon the two and the multiple, and does not recompose itself.

Here, there is no longer anyone able to walk the absolute spirit like a dog on a leash, strolling along the avenues of history. There is no longer any teleology. There is no longer anything that, apart from ourselves, can straighten out the way things work.

The second common element that these two positions develop is the imposing phenomenology of difference, which they both seek to interpret. Consequently, for both, the practice that subverts the human condition is, in the first instance, pushed towards *separatism*. 'Working class without allies': the workerist slogan declaims. Women rebel against the bourgeois institutions of patriarchal domination: this is how the initial feminine awareness of difference is organized polemically.

It is important to insist on this first common move of the two philosophies that interest us: obviously, this common feature is completely indistinct and formal from the standpoint of its contents (and it is no coincidence that these positions often clashed, each time patriarchal behaviours, induced by the wage-system itself, forced themselves arrogantly on proletarian families). However, this shared feature is no longer indistinct or formal when one considers that the process of separation, insisting on difference, will produce another deeper passage—a passage, an ontological turn that belongs to both these positions. A creative caesura.

But, some will cry out, these adventures of bodies and minds were well-known all over the world around 1968: why insist on the *Italian* specificity of these experiences?

An answer to this question can be given from two perspectives. A first unfolds from a cultural point of view. In the Italian desert, in this country that lacks a centre, unlike what happened in other NATO countries, the philosophies of difference developed in a purer form and did not need to express themselves through pre-existing paradigms. These movements constituted, so to speak, a real cultural and linguistic *epoché*. There was not much that could oppose them, except for the various modulations of the relationship of domination: the corporatist theories and practices of the industrial order, as well as the Catholic rules of good family life. It is therefore in the desert that these new and extremely strong flowers were born: it is in contrast to the deserted horizon, in the exotic, extremely potent prominence of their expression that the new forms of philosophical resistance and affirmation make themselves felt.

But there is another positive, constructive, and dynamic element that must be emphasized. For the first time, these philosophical 'differences' were unveiled in the biopolitical field (that is to say, they began to reveal the immediate political meaning of life itself). Now, this immediate biopolitical tension caused these differences to proliferate, to produce innovation. Over an exceedingly short period of time, Italy experienced the passage from the *separatist* affirmation of difference to its *constituent* affirmation. In fact, these diverse theories of difference did not simply represent a resistance to oppression and the oppressor; they were not entrenched in defensive positions, but became resistance that is productive; they showed that they were a manifold guerrilla movement. Here, there was no longer simply a theory, but a transformative practice. The practices embedded themselves in the junctures of social communication, threatened in micropolitical forms the major directives of capitalist and patriarchal command, carried out raids into knowledge and the universities, factories and workplaces, families and general social relations. Separation, understood from the standpoint of these two positions (which have never become theoretically blurred but have often been politically hybridized), turns into a *creative difference*. In Italy, this passage precedes those that, in different ways, will take place later elsewhere.

What is at issue here is really a caesura, a break, an ontological event. In this ubiquitous passage from separation to creative difference, from resistance to exodus, the movements and the consciousness of workers and/or women overcome the theme of the mere critique of the existent (a classical theme in the theories of organization of the modern era) and replace it with that of metamorphosis, of an inner and collective modification/transformation, both singular and ethical, led in the multitudes and by them. It is an exodus from this existence and from all its rules. This event will characterize the twenty years after 1968, and will increasingly deepen the subversive power of movements.

Here we come to a point where we can answer the previous objection: how different and how much more powerful are these Italic theories and practices of the subversive proletariat and feminine difference from the philosophical conceptions and the communal practices that derive from (and establish themselves in) the post-structuralist conceptions of difference—constructed, for instance, in France between Merleau-Ponty, Foucault, and Deleuze, between *Socialisme ou barbarie* and Luce Irigaray? Certainly, there are numerous kinships, but kinship does not here mean in any way filiation, because even in the rare cases when these positions originated from French theories, they then went on to live and flourish in wild *milieux*. They are products of the jungle... Indeed, from the outset, when the conceptions of difference developed into separatism, they moved from an ontological *irreducible*, which is immediate, forged in struggles, constructed in continuity. On the other hand, French philosophy only arrives at this ontological irreducible (be it the 'body without organs' or the 'production of subjectivity') at the end of its journey. Moreover, when it was a matter of constructing new horizons starting from this newly discovered field, French philosophy, at best, voiced wishes, constructed some hypotheses, rather than producing experiences and laboratories of life, rather than initiating an exodus. In Italy, the biopolitical field of difference has been explored in all its ethico-practical intensity, and the gaze of these practices has been fixed on what is *to-come* [*a-venire*]. In repression and in darkness, in the moments of that absurd Calvary that the Seventies and Eighties represented for the movement, a new light was born.

The thought of creative difference was also asserted against the philosophies of postmodernity. That completed and insignificant world (subsumed by capital) in which postmodern philosophers move is a world that shifts any possible critical space towards the outside, to its margins. On the world's fold, or on its limit, or where *zoé* opposes *bios*: here are the *extramoenia* deserts from where perhaps, or solely, resistance is possible—this is what the philosophers of the postmodern believe. Now, the practices of difference have opposed and refused these constructions of the postmodern, anticipating a longing for reconstruction based on the very affirmation of difference. The fact is that *difference* is *resistance*. The difference that is set out here is then placed as a break at the centre of the system of subsumption of social labour under capital and, in the order of reproduction, against the universal validity of feminine obedience to patriarchy—a resistance that breaks this horizon of domination not from the margins but from the centre, or better, which reconstructs a centre, a point which one can use as a lever in order to transform reality at the very heart of the system. A creative difference, an intense and radical exodus.

While re-valorizing the Italian philosophical scene beyond the cultural pages and academia, these positions of difference have become the seed for a new philosophy at a global level. As in the times of prophets, the philosopher needs to walk across the desert in order to express his thought. And again the prophet will not be listened to in his own country, but only outside it, yet, globally. In fact proletarian difference developed into intellectual power and opened onto a new social conception of labour and new forms of the production of value. Faced with this new reality, the structures of capitalist power [*potere*] had to be renewed; this is how the cataclysm in the global organization of power [*potere*] that we witness today originated. Beginning always from this proletarian difference, the new subject of the historical project of liberation constructs himself, as that multitude constituted by infinite singularities which will never again be subject to a sovereign command.

Yet again, it is feminine difference that is ultimately located at the centre of this radical modification of the philosophical horizon, because it has become a representation of biopolitical power and constitutes a real production of the social bond. There is, as it were, a second degree of creativity, what Spinoza identified in the advancing of the affective powers towards the creative condition of being, of corporeal *cupiditas* towards ontological *amor*, which is represented by feminine difference: we are dealing with a second degree of creativity that integrates and accomplishes the first difference, that of labour, expressing it as the original and general capacity for transformation possessed by bodies and the social whole. Genealogy includes and sublimates social labour.

Note that it is no longer Mario Tronti or Luisa Muraro who lead this revolution. Like all great authors, they have not bequeathed schools but rather lineages that operate on larger stages. The forms, modes, contents, and perspectives of difference left the world of seminars and workshops: they are in operation today and are developed and reinvented in the movements and the new social networks of productive cooperation. Difference has really stopped being separation: it has become creative and is beginning to produce the future. The exodus is hegemonic.

We can be proud that this happened in the Italian desert. Maybe now new resistance and new generations have their temple.

I am rereading my text. I have the impression of having given in, naïvely, to a certain *hubris*, of not wanting to take into account (as the *Angelus Novus* does) the horizon of destruction and death that still smolders behind us—perhaps, of not having thus accepted the sorrow that accompanies us no matter what. But is it really only a storm-wind that propels us? Conversely, doesn't the tragedy of our existence consist instead in subjective uncertainty and the risk of what is *to-come*? Doesn't the indubitable tragedy of our existing come somehow down to short-sightedness and unbearable fatigue in looking ahead, rather than to the feeling of defeat and the fierce incompletion that precedes us? Can't we oppose *cupiditas* to *hubris*? Deleuze once said: 'I hope to be the last thinker castrated by the history of philosophy'. I would like to paraphrase this: I want to be the last man castrated by the past, whose historical teleology I am incessantly the product of. But can the sense of difference and the event, together with the perception of the singularity that is constituting what is *to-come*, free me from the nightmare? This doubt is no less strong than the classical, Cartesian, metaphysical one, whose violent rationalistic *dispositif* of modernity we have had to endure. The doubt of non-truth is in fact a nightmare, but nightmare is a variation of dream. Where are we when we dwell between nightmare and dream? Between past and future? How can we, against all pessimism of reason, oblige the *Angelus Novus* to look forward, to settle the debt that it has contracted with history, and to overcome the constriction of the past?

Let's return to our point. The difference that shows itself as creative is the passage that leads the nightmare back to the dream, and the dream to a project (which is fully aware of the difficulty and limitation) of life. If difference is resistance, the dream can live its historical projection in a fully aware and conscious manner. If difference is a mode of life, it identifies the mode of life as productive. No one here is putting critical and transcendental action into question: but we should pity it, comprehend it in the radical aporia that gives rise to it, and which does not allow it to take root in the only natural and temporal difference that counts: that of power. As a matter of fact, in the theory of creative difference there is something like an extremely strong return to an origin that is not burdened with nightmares and repressive violences. This is not an illusion but the very thing which is here at stake. Difference does not become creative when it identifies itself with an origin (burdened by the past), but when it confounds itself with a power that is always new, open onto what is *to-come*. Difference destroys every determinate ontological foundation because it is the creative determination of an ontology of freedom.

No, neither Husserl nor Gentile nor Bergson support us here. Rather, we are aided by that strange, hard positivity of the only existing thought that is consistently immanentist and materialist, the one that we have revisited here. In the Italian twentieth century, this took the names we mentioned above. We thus have a paradoxical difference that is capable of positively producing the whole, bit by bit. A difference that knows how to develop into a network and to move from inside to outside, from the singular to the common, without solution of continuity, and vice versa. That knows how to be *res gesta* (after having destroyed the damned *historia*). It is with great respect for the story of the *Angelus Novus* that we ask it—following the rhythm of the practice of difference—to look forward. 'Another' world is possible, just as another place (that of difference) was possible in our Italian province.

Translated by Lorenzo Chiesa

3 FOUCAULT *DOCET*

Pier Aldo Rovatti

In spite of appearances, Antonio Negri's obscure pamphlet 'The Italian Difference' does not really lend itself to a polemical discussion. It must be taken for what it is, a *coup de théâtre* dictated—as the author himself confesses—by a rather ingenuous moment of *hubris*. At the end of the day, it is a *sparata*, as we say in Italian. Such a blast would intend to strike at the entirety of Italian contemporary thought (and with particular violence against so-called 'weak thought') in its capacity as a *philosophy of the master*; at the same time, it positively exempts from this treatment three names—the old Gramsci, and the new Mario Tronti, the workerist, and Luisa Muraro, the feminist—in their capacity as, it would seem, *philosophies that creatively resists the master by means of difference*. Everything else is a desert.

If there are no doubts about Gramsci, the two other names are—even for an Italian—quite unexpected. I wonder what those concerned by this bizarre ordering think about it (and then I ask myself: What status does he who arranges them arrogate to himself? Is he like the fourth man officiating at a football match?). Mario Tronti stood aside many years ago, avoiding the public scene; Luisa Muraro, whom I know very well, is on the contrary very present on the feminine front. She appears on television and even in glossy magazines without fear of becoming what Negri would call a 'game show hostess'. Along with these choices, we could produce many others, just as arbitrary and personal: this reminds me of the habit, quite in vogue in Italy, whereby everybody imagines himself to be the coach of the national football team and dictates his own line-up. So, I'll drop it.

As for Negri's intended targets, they revolve around the old motif, often used in reactionary terms, of the 'poverty' of Italian philosophy. I just want to say something about weak thought ('the vilest point' of the twentieth-century decline, as Negri delicately describes it), considering the fact that, at the beginning of the 1980s, I was its promoter together with Gianni Vattimo. Weak thought was an episode in the Italian philosophical debate that aroused considerable alarm in academia and whose effects (which also had significant international echoes) have yet to die out. These effects, which in part intersected with those of deconstruction, should induce some caution even in the worst-disposed of critics. I mean that, were he to exercise such caution, Negri would realize that what is at stake here is an issue of power [*potere*] that concerns the so-called metaphysical violence of philosophy, its administration of truth, and the elements of microgovernment that follow from it, beginning with the real privileges that exist in the institutional circles of research.

I think Negri is well aware that there is a front of struggle within philosophy, related to the very way in which the *scientificity* of concepts is understood and *knowledge as power* [*potere*] is used. Negri's sharp mind cannot overlook this Foucauldian inspiration of weak thought, unless he does so deliberately. As a matter of fact, his *very strong* thought could obviously fall into the critical horizon of weak thought itself.

I'm sorry to say this to a friend like Negri, but his pamphlet on the Italian difference is full of superficialities, that is, hurried verdicts which, as if wielding a machete, take the place of the reflection required by critical discourse. When sarcasm becomes the systematic shortcut for analysis, I doubt that philosophy remains (as Negri writes in the opening of his pamphlet) 'that critical activity that allows one to grasp one's time and orientate oneself in it'. I fully agree with this definition, to the extent that I'd like to bear it in mind when discussing some of the, so to speak, *serious* aspects that underlie Negri's text—since it is clear that something serious, and thus really discussible, both inspires it and makes it cohere.

My first observation concerns the emphasis on *national character*. The author of *Empire*, and other volumes that articulated its hypothesis, is the same author who, ever since his association with Mario Tronti at the beginnings of so-called 'workerism', has always stressed the multinational and ultimately global dimension of capitalism, and the correlative international and worldwide dimension of anti-capitalistic struggle. The subject in transformation in these struggles (in the sense that this subject is by now no longer the factory worker of the 1960s) does not identify himself with national characteristics, and, if he does so, he condemns himself to a theoretical and practical delay that corresponds with the worst vice of the 'Lefts' that Negri dislikes, insofar as they are inevitably retrograde vis-à-vis the *imperial* phase in which we live.

But in that case even an analysis of philosophical thought in specifically national terms will be similarly retrograde and, in a word, *anachronistic*. For instance, an evaluation of weak thought that did not take into account the international intellectual horizon, European and extra-European, in which it operates (Michel Foucault's discourse on power [*potere*], Jacques Derrida's deconstruction, Richard Rorty's ideas), would be meaningless. Just like other struggles, philosophical struggle, with its fronts of resistance, takes place on a global scene that far exceeds—in its practices—national vicissitudes.

What's more, Negri's own thought is clear proof of this: it is enough to observe the use he makes of a number of philosophical contributions to describe the condition of *empire*, and, in particular, the valorisation of theoretical strategies borrowed from Gilles Deleuze and Michel Foucault. In my view, it follows from this that the problem of an 'Italian difference' simply does not exist or is entirely artificial (and backward-looking). While what evidently remains important, and perhaps essential, is the *problem of difference*. It is with regard to this problem that we are called to provide critical clarification, one that acknowledges its genealogy (thus passing through Nietzsche and Heidegger); evaluates the legacy of a phase (basically, that of Deleuze), about which it is justifiable to ask whether difference has inclined towards a metaphysics of difference; and finally gives the right weight to Derrida's proposals (and thus his idea of deconstruction) which, as it is well-known, produced considerable insight into social practices, for instance into those of women. This is a problematic picture, one that needs to be explored with proper attention and circumspection. On the contrary, Negri simply assumes it at first blush, with excessive haste.

My second critical observation concerns Negri's overall approach. First, I note his insistence on the word *creativity*. It is difficult to disagree with him: behind this philosophical operator lies the couple Bergson-Deleuze, on which the group behind *aut aut*, the journal I edit (and which Negri knows well), has worked a lot in recent years. It is doubtless the case that Deleuze elaborates the idea of creative philosophy and conceptual *invention* with great originality. However, Negri applies it to a scenario that appears politically pregnant yet somewhat simplified. For him, it is as if there were only two levels: that of anti-capitalist struggles, and that of the theoretical tools suitable to represent and promote them. Any other term or mediation is excluded.

This scenario is a little bit magical, and actually cuts through differences and conflicts on both sides. It is by no means irrefutable that global struggles produce a uniform intelligence and, given that *diversity* appears to be a basic assumption with regard to this point, it is possible, and even desirable, that *different expressions of subjectivity* realize themselves, with valid points of individual specificity. Would the task of philosophy be that of unifying them in a strong thought of struggles? Answering 'yes' is problematic, to say the least. An answer that would guarantee such diversity using a thought that is sufficiently supple and open to the plurality of instances would seem more coherent. A thought that knows how to put itself on the line and renounce the haste of truth, in brief, a thought that is able to ward off its claims to hegemony.

All this leads us to the other dimension of the problem, itself characterized by a debate between positions, that is, by what could be called a struggle within philosophy. There is no trace of this in Negri's pamphlet, while, in my view, such a conflict of positions should be taken very seriously if one wishes to broaden one's perception of practices and form a microphysical picture of the balance of powers [*poteri*] in theoretical struggles. Foucault *docet*. Otherwise, at every turn, one runs the risk of taking refuge in positions exposed to dogmatism and, conversely, carrying out unproductive erasures, that is, throwing out the baby with the bath water.

The friend-enemy paradigm is useful only if it is the result of a search for identity, not the presupposition of a discourse. It seems banal to observe that if we speak of philosophy we first of all need to understand each other well regarding the status of philosophy and the so-called philosopher. Once we have agreed on the fact that philosophy must be a critical orientation vis-à-vis present reality, multiple foldings and differentiations appear on the scene, and there impose themselves just as many deconstructions of this very assumption which, at every turn, tends to congeal into an abstract presupposition. As Negri knows perfectly well, these foldings and differences are not the creation of isolated minds, uprooted from actual contexts, and they measure themselves up to the identity and status that the so-called philosopher assumes at any given time.

Such a 'philosopher' is never completely foreign to a disciplinary apparatus, to be understood as the historical disciplinarity of philosophy *qua* knowledge and as an institutional apparatus in which his practice of thought is always being produced. Without a critical clarification concerning the philosopher's stance both within and without discipline, we run the risk of turning him into a mythical figure and maybe, precisely for this reason, one who is very close to the logic of the master.

Translated by Lorenzo Chiesa

4 NIHILISM AS EMANCIPATION[1]

Gianni Vattimo

How can we speak about emancipation, that is, a process of liberation from constraints in the direction of greater freedom, autonomy, and possibility of choice, while associating it to concepts such as those of nihilism or hermeneutics? First of all, we should note that—as I had the occasion to show and illustrate in a number of books—the terms of nihilism and hermeneutics are here used as synonyms. Nihilism is understood in the sense inaugurally outlined by Nietzsche: the dissolution of all ultimate foundations, the awareness that—in the history of philosophy and Western culture in general—'God is dead' and 'the real world has become a fable'. Is this valid only for Western thought and culture? This first difficulty is not thematically discussed here; yet, Nietzsche—and Heidegger, and Marx before him, and even Hegel—teach us that the growing awareness that we think only within the ambit of Western culture is indeed part of such culture and its nihilism, since the very idea of a universal truth and a transcultural humanism (as for example in the doctrine of natural law or ultimate grounds) matured precisely within this given culture. When Western philosophy becomes aware of this, it becomes nihilistic; it takes note that its reasoning is always historico-culturally situated, that even the ideal of universality is 'comprehended' from a determinate point of view. But with this nihilism becomes hermeneutics: a thought that knows it can aim at the universal only by passing through dialogue, agreement, or *caritas*, if you like it (see my *Belief* and *After Christian-*

---

[1]. This short text was delivered at a conference in Ankara in the spring of 2006.

ity). '*Veritatem facientes in caritate*': translated into the terms of today's philosophy, this Pauline motto—which moreover echoes, maybe not from afar, the *aletheuein* of Aristotle's *Nichomachean Ethics*—means that truth is born in agreement and from agreement, and not vice versa, that we will reach agreement only when we have all discovered the same objective truth.

Emancipation is for us the meaning of nihilism proper if we read this Nietzschean term in the light of another crucial expression of the German philosopher: 'God is dead, and now we wish for many gods to live'. The dissolution of foundations (in which we can even recognize the moment of the passage from modernity to post-modernity—see my *The End of Modernity*) is that which frees us—once again, with a profound echo of the Gospel 'The truth shall make you free'. Does this mean that 'knowing how things "really" are will free you'—finally discovering Pythagoras' theorem? The necessary geometrical order of the world? Einstein's relativity? No. Rather, it means that 'truth is only that which frees you'; truth is thus first of all the 'discovery' that there are no ultimate foundations before which our freedom should stop, which is instead what authorities of every kind that want to rule precisely in the name of these ultimate foundations have always sought to make us believe. Hermeneutics is the thinking of accomplished nihilism, the thinking that aims at a reconstruction of rationality after the death of God, in opposition to any drift towards negative nihilism, that is, towards the desperation of those who continue to grieve because 'there is no more religion'.

It is clear that all this has significant implications for the way one conceives of ethics, law, and politics. After the death of God, will it still be possible to talk about moral imperatives, laws that are not founded arbitrarily, and an emancipatory horizon of politics? My work does not delude itself into believing that it gives exhaustive answers to these questions; but neither does it limit itself to echoing them rhetorically—this is what much contemporary tragicism [*tragicismo*] does, exhausting itself in the rhetorical emphasis of the problematicity of the human condition, often in order to prepare a 'leap of faith' (which then becomes a leap into pure irrationality and the subsequent defection to the dogmatic authoritarianism of churches, central committees, and charismatic leaders), or, at other times, in order to maintain itself in the pure and simple awareness that 'there is no solution', with the tacit pretension that, socratically, knowing not to know is always better (Nietzsche was right in unmasking the optimistic rationalism of such a demeanor).

The hermeneutic exit from tragic and negative nihilism naturally also entails the retrieval of many of its aspects; one should say, with Nietzsche, that one cannot build without destroying. Or even, more realistically, one should say that the mother of all metaphysical authoritarianisms is always pregnant, hence the task of secularization—that is, the unmasking of the sacredness of any absolute, any ultimate truth—is far from having become outdated. Politics, law, and social life continually bear witness to this claim, not only in Italy, where the Catholic Church continues to (demand to) impose unreasonable limits on the state's laws (think of civil unions, research on embryos, and euthanasia), but now also in international politics, where American dominance masked as democratic humanitarianism threatens to impose a kind of universal state of police which is 'legitimized' by an (alleged) respect for human rights, or those that the empire considers such. Won't the new Napoleon instigate some new 'romantic' rebellion of nations—of cultures, of 'people' (with all the reservations that should be induced by these terms)—against the armed *pax Americana*?

Trying to measure up to such problems—albeit in a very theoretical way—hermeneutics thus inherits first of all much of the critical and 'destructive' contents of tragic nihilism. But hermeneutics also harbours two openings towards constructiveness. First of all, the death of God does not claim to be a finally achieved truth, on the basis of which one could dogmatically found some natural law of atheism, of the 'unfounded' world, or of some Nazi-type *Übermensch*. The constructive nihilism of hermeneutics does not only have to defend itself from the neurotic return of authoritarianisms, but also from the metaphysical sclerosis of antifoundationalism (for instance, the latter easily goes hand in hand with the imposition of freedom and democracy by means of armed interventions against what President Bush named 'rogue states'—these usually are such, but it is not Bush or the United Nations transformed into an ethical court of law that can pass judgement on them). To all these distortions of nihilism, hermeneutics first and foremost opposes the very principle of the plurality of interpretations, that is, the principle of the respect for everyone's freedom of choice. Certainly, this is not much more than Habermas's communicative rationality; but the latter is here stripped of the remnants of metaphysical rationalism that still invalidate it—such a theory, with its idealisation of a knowledge freed from opacities and ultimately modelled upon the scientific method, always runs the risk of legitimising a future world dominated by 'experts' of various kinds. The critical weapons of negative nihilism thus remain decisive for the constructiveness of hermeneutics. Attempting to shape laws, constitutions, and ordinary political measures, according to the idea of a progressive liberation of norms and rules from any alleged 'natural' limit (i.e. one that is manifest only to those who possess power) can already constitute a positive political project. Recall that, already many years ago, a theoretician close to Habermas like Karl Otto Apel (see his *Transformation der Philosophie*) even accounted for the fight against world hunger on the basis of the respect for the equal rights of our interlocutor, which is imposed on us by any use of language, on pain of a performative contradiction. That is to say: even when I speak only to myself I have to respect some rules; I am responsible for such respect before any interlocutor, which means that I grant my same rights to any interlocutor; but then I must also positively guarantee him the conditions for the exercise of these rights, and consequently the human conditions of survival. Now, the hermeneutic (and 'nihilistic') ideal of founding every law and social behaviour on the respect of everyone's freedom and not on allegedly objective or 'natural' norms implies positive consequences that are much broader than those that Apel indicated in his work of the 1960s—after all without giving them an explicitly programmatic development. For instance, peace—even when it is not understood too theologically as the 'tranquillity of order', according to Augustine's phrase which the Catholic Church has used to justify its worst silences on Fascism and Nazism—is a basic human right whose topicality and problematicity has sadly come to the fore recently. Isn't the reforming of constitutions and the drafting of laws that take into account rights like this also the basis of a positive political programme? At the end of the day, this is what marks the (necessary) passage from liberalism to democracy and, for us, socialism; in order really to achieve the rights of freedom preached by liberalism, we should not let things take place 'according to their own principles', as for example, in the laws of the market (there is an unacceptable 'naturalism' in Adam Smith!). Rather, we must build conditions of equality that, indeed, are not given 'naturally'.

If we wish to summarize in a few words the meaning of a nihilistic hermeneutics—one that is, after all, an entirely open enterprise—what I myself see in it at this moment is a confirmation of Heidegger's thesis on being as 'event', and not as a stable structure given once and for all (what Heidegger calls 'metaphysics'). An event that is possible only on condition that being 'is not', or is no longer—on condition that God is dead and that the eternal structures of values have been unveiled as a lie. Only on condition of traversing the experience of nihilism understood in this way is it possible to plan a society where freedom will not be an empty term: truth is always 'to be made', and thus values are always to be invented anew. It is in nihilism thought in this way that equality finally establishes itself, and what Richard Rorty calls solidarity becomes possible—or better necessary—for life, the only possible basis for a truth that does not claim to evade the historical conditions in which existence is always 'thrown'.

Translated by Lorenzo Chiesa

5 COMMUNITY AND NIHILISM

Roberto Esposito

1. What is the relationship between the terms 'community' and 'nihilism'? The answer that comes from the various philosophies of community—but also from a widespread interpretation of nihilism—goes in the direction of a radical opposition. Nihilism and community are not just in a relation of alterity, but in one of open contrast, which does not admit points of contact or areas of overlap. They reciprocally exclude each other: where one is present—or, when one is present—the other is absent, and vice versa. Whether the opposition is located on the synchronic level or along a diachronic trajectory, what matters is the clarity of the alternative between two poles that seem to acquire a meaning precisely from their irreducibility. Nihilism—in its most distinguishing connotations of artificiality, anomie, and senselessness—is perceived as that which has made community impossible, or even unthinkable. On the other hand, community has always interpreted itself as what resists, restrains, and contrasts the nihilistic drift. This is basically the role assigned to community by the communal [*comuniali*], communitarian, and communicative conceptions which, for more than a century, have regarded it as the only barrier against the devastating power of nothingness which pervades modern society. What changes, with regard to this scenario, is the order of succession that is attributed at each turn to community and modern society, not their rigidly dichotomous character. If Ferdinand Tönnies put community before society—according to a genealogy which was then appropriated by all the philosophies of decline, betrayal, and loss originating both from the Right and the Left at the turn of the twentieth century—

contemporary neo-communitarians across the Atlantic reverse the stages of the dichotomy, yet without questioning its basic structure. It is community—or better, the particular communities into which the Tönniesian archetype has been fragmented—that follows modern society in a phase marked by the crisis of the state paradigm and the proliferation of multicultural conflict. In this case, community is no longer understood as a residual phenomenon with regard to the sociocultural forms assumed by modernity, but rather as an objection to the insufficiency of the latter's individualistic-universalistic model: it is the very society of individuals, the destroyer of the ancient organic community, that now generates new communitarian forms as a posthumous reaction to its own inner entropy. Even from this perspective, what re-emerges is the reciprocal exclusion of community and nihilism: community advances or withdraws, expands or contracts itself, on the basis of the space which has not yet been 'colonized' by nihilism. When Habermas opposes a communicative to a strategic rationality, he remains within the same interpretative paradigm, with an additional, defensive emphasis: the 'unlimited community of communication' constitutes, at the same time, the point of resistance and the reserve of meaning in face of the increasing intrusiveness of technology. The fact that community is understood as a transcendental a priori—rather than a factual one, like in the more rudimentary approach of the neo-communitarians—does not change its basic hermeneutic frame. Even in this case, community, considered as a possibility if not a reality, is understood as the borderline and the wall that contains the advance of nihilism. It is seen as something full—a substance, a promise, a value—that does not let itself be emptied out by the vortex of nothingness. It is another configuration of the battle between the 'thing' and the 'nothing' that functions as a presupposition for the entire tradition we are examining: against the explosion—or the implosion—of the nothing, community holds back the reality of the thing: or rather, it is the very thing that opposes its own annihilation.

2. But is this an acceptable presupposition? Is it not itself precisely what hinders any thought of a community that would be worthy of our age—which is indeed the age of completed nihilism? If we assumed this presupposition as such, we would necessarily

be obliged to choose between two hypotheses which are equally unacceptable. That is, we would find ourselves either negating the structurally nihilistic attitude of the present age, or excluding the question of community from our horizon of relevance. In order to speak about community in terms that are not simply nostalgic, we would be left with the possibility of circumscribing nihilism to an aspect, a particular moment, of our experience. We could consider it as a 'fixed term' phenomenon, bound to dissolve itself or at least regress at a certain point. Or we could even understand it like a disease which has attacked only some organs of an otherwise healthy body. Yet this kind of reductive argument goes against all evidence, which shows that nihilism is not an interlude or a specific situation, but rather the basic tendency of modern society, which has today achieved its utmost expression. But what does this mean? The only way to get our head around the question without renouncing any of its terms requires bringing together community and nihilism in a single argument. Even more, we should not see the completion of nihilism as an insurmountable obstacle, but as an opportunity to elaborate a new thinking of community. Obviously, this does not mean that community and nihilism turn out to be identical, or even just symmetrical; that they are to be located on the same level or along the same trajectory. Rather, it means that they intersect at a point that the two cannot disregard because, in different ways, it is constitutive of them both. This point—which is unperceived, repressed, and neutralized by current communitarian philosophies—can be regarded as the 'nothing'. It is this that community and nihilism have in common in a way that has so far remained mostly unexplored.

But in what sense? We leave aside for the time being the (far from simple) question about the relation between the nothing and nihilism—yet we shall return to it in a short time. Let us focus on community. We have seen how it has traditionally been opposed to nihilism as if it were the thing itself; and even how its definition is one with such an opposition: community would not just be different from the nothing and irreducible to it, but it would also coincide with its explicit opposite—a 'whole' entirely filled by itself. Now, I believe that this is precisely the standpoint that should not only be problematized, but even reversed: community is not the place of the opposition of the thing and the nothing, but that of their superimposition. I have attempted to account for this claim by means of an analysis, both etymological and philosophical, of the term *communitas*, starting from the term *munus*, from which it derives.[1] The conclusive result of this investigation is community's categorical distance from any idea of property collectively owned by a group of individuals—or even from their belonging to a common identity. According to the original value of this concept, what the members of *communitas* share—this is precisely the complex, but pregnant meaning of *munus*—is rather the expropriation of their substance which is not limited to their 'having', but involves and draws on their very 'being subjects'. Here, my argument unfolds in a way that shifts it from the more traditional field of anthropology, or of political philosophy, to the (more radical) field of ontology: the fact that community is not linked to a surplus, but a deficit, of subjectivity, means that its members are no longer identical to themselves, but structurally exposed to a tendency that leads them to break their individual limits and face up to their 'outside'. From this point of view—which breaks any continuity between what is 'common' and what is 'one's own' [*proprio*], linking it rather to what is not one's own [*improprio*]—the figure of the other returns to centre stage. If the subject of the community is no longer the 'same', he will necessarily be 'other'. He will not be another subject, but a sequence of alterations that never coalesce into a new identity.

3. But if community is always the community of others and never of oneself, this means that its presence is structurally inhabited by an absence—of subjectivity, identity, and property. It means that it is not a 'thing'—or, it is a thing defined precisely by its 'not'. A 'non-thing'. Now, how should we understand such 'not'? And how does it relate to the thing it inheres to? What is for certain is that it does not relate to it in the sense of a pure negation. The nothing-in-common is not the opposite of an entity, but rather something that corresponds and co-belongs to it in a very intense way. Yet we should not misunderstand the very meaning of this correspondence, or co-belonging. The nothing of *communitas* should not be interpreted as what *communitas* is not yet

---

1. See Roberto Esposito, *Communitas. Origine e destino della comunità*, Turin, Einaudi, 1998.

able to be; as the negative moment of a contradiction bound to be solved dialectically in the identity of opposites. But neither should it be interpreted as the hiding place in which the thing withdraws since it cannot unveil itself in the fullness of a pure presence. As a matter of fact, in both cases, the nothing of *communitas* would not continue to be the nothing of the thing, but it would be transformed into something different which the thing would relate to in the modes of teleology or presupposition. It would be the thing's past or its future, not its bare present—that which it is and is not other from it. In short, the nothing is not the precondition or the outcome of the community—the presupposition that frees it for its 'real' possibility—but rather its only way of being. In other words, community is not proscribed, obscured, or veiled by the nothing: it is constituted by it. This simply means that community is not an entity, nor a collective subject, nor a group of subjects. It is the relation that makes them no longer be individual subjects, since it interrupts their identity with a bar that passes through them and thus changes them. It is the 'with', the 'between', the threshold on which they cross in a contact that relates them to others to the very extent that it separates them from themselves.

We could say that community is not the *inter* of the *esse*, but the *esse* as *inter*; not a relationship that shapes being, but being itself as a relationship. This is an important distinction since it gives us back in the clearest possible way the superimposition of being with the nothing: the being of community is the gap, the spacing that relates us to others in a common non-belonging, a loss of what is one's own which never manages to be added up into a common good. Only lack is common, not possession, property and appropriation. The fact that the term *munus* is understood by the Latins only as the gift given, and never as the gift received—which is instead rendered by the word *donum*—means that it is a principle that lacks 'remuneration'. It means that the leak of subjective substance which it determines stays there—it cannot be filled in, cured, or cicatrized; that its opening cannot be closed by any filling in [*risarcitura*], or compensation [*risarcimento*], if it is to remain really condivided [*condivisa*], or shared. In the concept of 'condivision' the 'con', or 'with', is indeed associated with 'division'. The limit it alludes to is that which unites, not in the mode of convergence, conversion, or confusion, but rather in that of divergence, diversion, and diffusion. The direction here is always from the inside to the outside, and never from the outside to the inside. Community is the exteriorisation of the inside. For this—given that it is opposed to the idea of internalisation, not to mention that of internment—the *inter* of community can only link exteriorities or 'leakages', subjects who face up to their outside. This movement of decentralisation can be recognized in the very idea of 'partition'—which refers to both 'condivision' and 'departure': community is never a place of arrival, but one of departure. It is even the very departure towards what does not and will never belong to us. Therefore *communitas* is far from producing effects of commonness [*comunanza*], association [*accomunamento*], or communion. It does not warm us up, or protect us. On the contrary, it exposes the subject to the most radical risk: the risk of losing together with his individuality also the boundaries that guarantee the fact that he is intangible for the other. The risk of suddenly slipping into the nothing of the thing.

4. It is with reference to this nothing that we must address the question of nihilism—in a way that is not only able to grasp the connection, but also the distinction of levels on which it is based. What I mean to say is that nihilism is not the expression, but the suppression of the nothing-in-common. Certainly, nihilism has to do with the nothing, but precisely in the guise of its annihilation. Nihilism is not the nothing of the thing, but that of the thing's nothing. It is a nothing squared: the nothing multiplied and simultaneously swallowed up by the nothing. This means that we should identify at least two meanings, or levels, of the nothing, which must be kept separate in spite of and within their apparent coincidence. While the first level is, as we have seen, that of a relationship—the gap, or the spacing, that makes the being-in-common a relation, not an entity—the second is, on the other hand, that of its dissolution: the dissolution of the relationship in the absoluteness of the without-relation.

If we look at Hobbes's absolutism from this perspective, the stages of such a 'solution' assume an extraordinary clarity. The fact that Hobbes inaugurates modern political nihilism should not simply be understood in the sense that he 'discovers' the nothingness of substance of a world freed from the metaphysical constraint of any transcendent *veritas*; Hobbes rather 'covers' this nothingness of substance again with another, more powerful, nothingness, which has precisely the function of annihilating the potentially dissolutive effects of the first. Similarly, the *pointe* of his political philosophy lies in the invention of a new origin aimed at damming up—and turning into an ordering compulsion—the original nothing, the absence of origin, of *communitas*. Obviously, such a contradictory strategy of neutralisation—emptying the natural void by means of an artificial void created *ex nihilo*—is derived from an altogether negative, and even catastrophic, interpretation of the principle of condivision, the initial sharing of being. It is precisely the inevitable negativity attributed to the original community that justifies a sovereign order—the Leviathan State—able to pre-emptively immunize itself from its intolerable *munus*. In order for this operation to be successful—that is, to be logically rational in spite of its very high cost in terms of sacrifice and renunciation—it is not only necessary that such common *munus* be deprived of its character as donative excess in favour of its character as defect, but also that this defect as lack—in the neutral sense of the Latin *delinquere*—be understood in terms of a real 'delict' [*delitto*], a crime, or even a unstoppable chain of potential crimes.

It is this radically forced interpretation—which turns the nothing-in-common into the community of crime—that determines the obliteration of *communitas* in favour of a political form founded upon the emptying of any relation that is external to the vertical relation between individuals and the sovereign, and consequently upon dissociation itself. Having started off from the need to protect the thing from the nothing that appears to threaten it, Hobbes thus ends up annihilating not only the nothing, but the thing itself; he sacrifices to the interest of the individual not only the *inter* of the *esse*, but also the *esse* of the *inter*. All the modern answers that have been given to the 'Hobbesian problem of order' in the course of centuries—in decisionist, functionalist, and systemic guises—run the risk of remaining caught in this vicious circle: the only possible way to contain the dangers that are inherent to the original lack [*carenza*] of man as animal seems to be the construction of an artificial prosthesis—the barrier of institutions—able to protect him from the potentially destructive contact with his fellow men. Yet, assuming a prosthesis, that is, a non-organ, a lacking organ, as a form of social mediation means facing the void with a void that is even more radical, since, from the beginning, it is seized and produced by the absence that it should compensate for. The very principle of representation, understood as the formal device aimed at giving presence to someone who is absent, only reproduces and strengthens that void insofar as it is not able to conceptualize its primordial character, which is not derived from anything. In other words, the principle of representation is not able to grasp that the nothing that it should compensate for is not a loss of substance, foundation, or value, which suddenly dissolved a previous order, but the very character of our being-in-common. Not wanting or knowing how to dig deeper into the nothing of the relation, modern nihilism finds itself being handed over to the nothing of the absolute, the absolute nothing.

5. The modern philosophy of community attempts to elude the absolute nothing through an option that is both similar and opposite to the one I have just described; however, it ends up falling back into the very nihilism it would like to fight against. In this case, it is the thing that is made absolute, rather than the nothing. But what does making the thing absolute mean, if not annihilating—and hence once again strengthening—the nothing itself? This strategy no longer empties, but, on the contrary, fills in the void which is determined, and even constituted, by the primordial *munus*. Beginning with Rousseau and up to contemporary communitarianism, what appears as an alternative option turns out to be the specular reverse of Hobbesian immunisation, with which it shares both the subjectivist lexicon and the particularistic outcome—this time applied to a collectivity as a whole, not the individual. What is missing in both cases—crushed by the overlapping of the individual with the collective—is relation itself, understood as a modality at the same time singular and plural of existence. In the first case, relation is annihilated by the absoluteness that separates individuals; in the second, by their fusion in a single subject closed within his self-identity. If we take the Rousseauian community of Clarens as a model of such an—infinitely reproduced—self-identification, we can detect in it *in vitro* all of its defining characters: from the reciprocal incorporation of its members to the perfect self-sufficiency of the

whole they give rise to, to the inevitable opposition that results from it with regard to its outside. The outside as such is incompatible with a community that is so folded towards its inside that it institutes among its members a transparency without opacity, an immediateness without mediations, which constantly reduces each member to another who is no longer such since he is pre-emptively identified with the first. The fact that Rousseau does not prefigure—and actually constantly denies—the possibility of translating such *communauté de coeur* into some form of political democracy does not eliminate the power of mythological suggestion that it has exercised not only on the entire Romantic tradition, but also, in different ways, on the ideal type of the organic *Gemeinschaft*—itself founded on the generality of an essential will which has precedence over that of its individual members.

But there is something else that pertains more specifically to this unwittingly nihilistic relapse of the opposition of community to the nihilism of modern society—to which community not only shows itself to be fully adherent, but also strictly functional as its mere reverse. Each time that the lack of sense of the individualistic paradigm has been opposed to the surplus of sense of a community filled by its own collective essence, the consequences have been destructive: first for the internal, or external, enemies against whom the community was established, and eventually for the community itself. This obviously applies in the first place to the totalitarian experiments which have stained with blood the first half of the last century, but also, in a different and less devastating way, to all forms of 'fatherland', 'motherland', or 'brotherland' [*fratria*] which have gathered crowds of followers, patriots, and brothers around a model inevitably centred on a *koine*. The reason of this tragic compulsion to repeat—which does not seem to be on the wane—lies in the fact that when the thing fills itself to the brim with its own substance, it runs the risk of exploding or imploding under its own weight. This happens as soon as the subjects gathered in the communal [*comuniale*] bond identify the access to their condition of possibility in the re-appropriation of their own common essence. The latter, in turn, appears to shape itself as the fullness of a lost origin, which would be for this reason retrievable in the internalisation of a temporarily exteriosized existence. In this way, it is assumed that it is possible, and even necessary, to elide—or fill in—the void of essence that constitutes the *ex* of *exsistentia*—its not being its own since it is 'common'. It is only in this way—by means of the abolition of its nothing—that the thing can finally be realized. Yet, the (necessarily phantasmatic) realisation of the thing is, as a matter of fact, the aim of totalitarianism; the absolute lack of differentiation that ends up suppressing not only its own object, but the very subject that puts it into effect. The thing can only be appropriated in its destruction. It cannot be retrieved for the simple reason that it was never lost: what appears to be lost is only the nothing that constitutes it in its common dimension.

6. The first thinker who looked for the community in the nothing of the thing was Heidegger. Although it is impossible to retrace here the complex trajectory of the interrogation about the thing that unfolded throughout his work, it is necessary to focus on the 1950 paper titled 'The Thing' (*Das Ding*). Such trajectory seems to culminate in this paper; even more crucially, the 'thing'—which is elsewhere addressed in its aesthetic, logical, or historical aspects—is here brought back to its common essence. This expression needs to be understood in a twofold way. First, in the sense that Heidegger summons up the most modest, ordinary, and down-to-earth things—in this text, the jug. But also in the sense that this modesty looks after the empty point in which the thing recovers its least expected meaning, as Heidegger had already argued in *The Origin of the Work of Art*: 'The unpretentious thing evades thought most stubbornly. Or can it be that this self-refusal of the mere thing [...] belongs precisely to the essence of the thing?'[2] The lecture on 'The Thing' is devoted precisely to the definition of this essence—'the thingness of the thing'. This does not amount to the objectivity in which we represent the thing; or to the production from which the (produced) thing seems to 'originate'. And so? It is precisely here that the example of the jug is helpful—but also that of the other 'things' Heidegger refers to in the essays of those years, such as the tree, the bridge, and the threshold. What characteristic element links them all? Basically, it is the void. The void is the essence of these things, as well as of all things in general. This is the case with the jug—which is literally gathered together around a void and

---

2. Martin Heidegger, 'The Origin of the Work of Art', in David Farrell Krell (ed.), *Basic Writings*, London, Routledge and Kegan Paul, 1978, p. 161.

is, in the last instance, formed by it: 'When we fill the jug, the pouring that fills it flows into the empty jug. The emptiness, the void, is what does the vessel's holding. The empty space, this nothing of the jug [*Die Leere, dieses Nichts am Krug*], is what the jug is as the holding vessel'.[3] The essence of the thing is therefore its nothingness, to the extent that outside of the perspective this opens, the thing loses its most intimate nature, to the point of disappearing—or, like Heidegger has it, to the point of being annihilated. As soon as we forget about its essence 'in truth, the thing as thing remains proscribed, nil, and in that sense annihilated [*In Wahrheit bleibt jedoch das Ding als Ding verwehrt, nichtig und in solchem Sinne vernichtet*]'.[4]

All this may seem to be paradoxical: the thing is annihilated if we do not grasp fully its essential character. Yet, as we have just seen, this essential character lies in nothing else than its void. It is the forgetting of this nothingness—this void—that hands the thing over to a scientistic [*scientista*], productivist, and nihilistic point of view which nullifies it. Even in this case, we find ourselves obliged to establish a distinction between two kinds of 'nothingness': the first gives us back the thing in its deep reality, while the second removes it from us. Or better still, nullifying the first nothingness, the second nullifies the thing itself that is constituted by it. Some lines later, Heidegger gives us the key to this apparent paradox: the nothingness that saves the thing from nothingness—to the extent that it essentially constitutes it as thing—is the nothingness of the *munus*, the offer that reverses the inside into the outside: 'To pour from the jug is to give [*schenken*]'.[5] Not only this, but this nothingness is the nothingness of the 'common' *munus* insofar as it gives itself in the gathering and as a gathering: 'The nature of the holding void is gathered in the giving'.[6] To this end, Heidegger recalls the old German words *thing* and *dinc* in their original meaning of 're-union'. The giving expressed by the void of the jug is also and above all a gathering. What is it that gathers together the void of the thing by offering it? Heidegger adds at this point the motif

---

3. Martin Heidegger, 'The Thing', *Poetry, Language, Thought*, trans. Albert Hofstadter, London, Harper Perennial, 1976, p. 169.
4. Heidegger, 'The Thing', pp. 170-1.
5. Heidegger, 'The Thing', p. 172.
6. Heidegger, 'The Thing', p. 172.

of the 'fourfold', that is to say, the relation between the earth and the sky, mortals and divinities. But what we should focus on is the relation as such: the nothing that it puts in common is the community of the nothing as the essence of the thing. Is it not precisely this—the pure relation—that constitutes the common element of all the things mentioned above: the tree that links the earth to the sky, the bridge that connects two banks, the threshold that joins the inside with the outside? Just as is the case with *communitas*, is this not a unity in distance and of distance; of a distance that unites or a separation that brings near? And what is, in the end, nihilism if not an abolition of distance—of the nothingness of the thing—that makes any nearness impossible? 'The failure of nearness [*das Ausbleiben der Nähe*] to materialize in consequence of the abolition of all distances has brought the distanceless to dominance. In the default of nearness the thing remains annihilated as a thing in our sense'.[7]

7. The only author who tackled the question opened by Heidegger—that of the relation between community and the nothing in the time of completed nihilism—is Georges Bataille: "Communication' cannot proceed from one full and intact being to another. It requires beings whose being in themselves is *risked*, placed at the limit of death and nothingness [*néant*]'.[8] This passage refers back to a short text entitled 'Nothingness, Transcendence, Immanence' in which nothingness is defined as 'the limit of a being' beyond which this being 'no longer exists, no longer is. For us, that nonbeing is filled with meaning: I know I can be reduced to nothing [*Ce non-être est pour nous plein de sens: je sais qu'on peut m'anéantir*]'.[9] Why is the possibility of being annihilated filled with meaning—and even amounts to the only workable meaning at a time when every other meaning seems to be waning? This question leads us to both Bataille's interpretation of nihilism and the point at which it crosses aporetically the inhabitable place of community. For Bataille, nihilism is not a flight of sense—or from sense—but rather its closure within a homogeneous and completed conception of being. There aren't

---

7. Heidegger, 'The Thing', p. 181.
8. Georges Bataille, *On Nietzsche*, trans. Bruce Boone, London, Athlone Press, 1992, p. 19 (my translation).
9. Bataille, *On Nietzsche*, p. 177.

other instances in which nihilism is less reducible to what threatens to empty the thing. On the contrary, nihilism is what clogs the thing in a fullness without cracks or fissures. In short, nihilism should not be looked for on the side of the lack, but on that of the subtraction of lack. It is the lack of lack—its repression or filling in. It is what subtracts us from our otherness blocking us inside ourselves; what makes that 'us' into a series of completed individuals who are turned towards their inside, fully resolved in themselves:

> Boredom then discloses the nothingness of self-enclosure [*le néant de l'être enfermé sur lui-même*]. When a separate being stops communicating, it withers. It wastes away, (obscurely) feeling that *by itself it doesn't exist*. Unproductive and unattractive, such inner nothingness repels us. It brings about a fall into restless boredom, and boredom transfers the restlessness from inner nothingness to outer nothingness—or anguish.[10]

What emerges here clearly is the twofold level of the semantics of nothingness and, at the same time, the movement Bataille carries out from the first to the second level—from the nothingness of the individual, of what is one's own, the inside, to the nothing-in-common of the outside. This second nothing is also a nothing, but it is the nothing that tears us away from the absolute nothing—the nothing of the absolute—since it is the nothing of relation. Man is structurally exposed to—but we should also say: constituted by—the paradoxical condition of being able to avoid annihilation by implosion only running the risk of annihilating himself by explosion: 'With temptation, if I can put it in this way, being is crushed by the twin pincers of nothingness. By not communicating, it is annihilated into the emptiness of an isolated life. By communicating it likewise risks being destroyed'.[11]

The fact that Bataille—here as elsewhere—speaks of 'being' alluding to our existence should not be interpreted only as a terminological imprecision due to the non-professional philosophical character of his thought, but as the intentional effect of an overlapping between anthropology and ontology within the common figure of lack, or, more precisely, the ripping [*déchirure*]. Indeed, it is true that we are able to face up to the being that lies

---

10. Bataille, *On Nietzsche*, p. 23.
11. Bataille, *On Nietzsche*, p. 24.

outside our boundaries only if we break them—and even identify ourselves with such a rupture. But this is due to the fact that being is also primordially lacking with regard to itself, since the ground of things does not amount to a substance but a primordial opening. We access this ground—this gap—in the limit-experiences that take us away from ourselves, from the mastery of our existence. Yet these experiences are nothing else than the anthropological effect (or the subjective dimension) of the void of being that originates them: a big hole made by several holes that alternately open themselves inside it. In this sense, we could well say that man is the wound of a being that is in its turn always-already wounded. This means that when we speak of the being-in-common, the 'communal' ['*comuniale*'], as a continuum into which every existence that has broken its own individual boundaries falls back, we should not understand this continuum as a homogeneous whole—this is precisely the nihilistic perspective. Nor should we understand it as being in the strict sense of the word—or as what is Other from being. We should rather understand it as a vortex—the common *munus*—in which the continuum is one with what is discontinuous, and being is one with not-being. This is the reason why the 'greatest' communication does not look like an addition or a multiplication, but rather like a subtraction. It does not take place in the passage between the one and the other, but in that between the other of the one and the other of the other:

> The beyond of my being is first of all nothingness. This is the absence I discern in laceration and in painful feelings of lack: It reveals the presence of another person. Such a presence, however, is fully disclosed only when the *other* similarly leans over the edge of nothingness or falls into it (dies). 'Communication' only takes place *between two people who risk themselves*, each lacerated and suspended, perched atop a common nothingness [*l'un et l'autre penchés au-dessus de leur néant*].[12]

8. We could well say that, with Heidegger and Bataille, twentieth-century thought on community reaches its point of maximum intensity and, at the same time, its outermost limit. This is not due to the fact that twentieth-century thought experiences

---

12. Bataille, *On Nietzsche*, pp. 20-1.

in their philosophies several relapses in a mythical and regressive direction; or because it is not possible to register—around and after these two authors—elaborations, developments, and new intuitions which, in different ways and with different inflections, refer back to the question of the *cum*: the writings—and lives—of Simon Weil, Dietrich Bonhoeffer, Jan Pato ka, Robert Antelme, Osip Mandelstam, and Paul Celan bear witness to the opposite. Rather, this is due to the fact that even these thinkers could think community only starting from the problem posed, and never solved, by Heidegger and Bataille. It is for the same reason that all that separates us from them—the philosophy, sociology, and political studies of the second half of the twentieth century—remains forgetful of the question of community. Or, worse, it contributes to the distortion of community whenever it reduces it to the defence of new particularisms. Only in the last few years, this drift—experienced and produced by all the ongoing debates on individualism and communitarianism—has been countered, especially in France and Italy, by an attempt to launch a new philosophical reflection on community that starts exactly at the point where the previous one was interrupted in the mid of the twentieth century (see Esposito 1998; Agamben 1993; Nancy 1991; Blanchot 1984).[13] The necessary reference to Heidegger and Bataille that characterizes this reflection is accompanied, however, by the clear awareness that we live with the inevitable exhaustion of their lexicon, that is, in a condition—both material and spiritual—which they could not know fully.

Once again, I am alluding to nihilism, and more precisely to the further acceleration that took place within its uninterrupted 'completion' during the last decades of the twentieth century. It is perhaps precisely this acceleration that allows—but also imposes on us—a recommencement of the thought on community in a direction which Heidegger and Bataille could only guess, but not thematize. What direction? Without presuming to offer an exhaustive answer to what is the question of our time, it is inevitable to take another look at the figure of the 'nothing'. Nancy, the

---

13. See: Giorgio Agamben, *The Coming Community*, trans. Michael Hardt, Minneapolis, University of Minnesota Press, 1993, Maurice Blanchot, *La communauté inavouable*, Paris, Minuit, 1984, Esposito, *Communitas*, Jean-Luc Nancy, *The Inoperative Community*, trans. Lisa Garbus, Peter Connor, Michael Holland, and Simona Sawhney, Minneapolis, University of Minnesota Press, 1991.

contemporary author who, more than any other, has the merit of having made a breach in the closure of the thought on community, writes the following: 'The question is rather to know how to conceive of the "nothing" itself. Either it is the void of truth, or it is nothing else than the world itself and the meaning of being-in-the-world'.[14] How should we understand this alternative, and is it really an alternative? To this end, we could observe that, from a certain point of view, it is precisely the absence of community—and even its desertification—that shows us its necessity as what we lack, and even as our own lack; as a void that does not ask to be filled in by new or ancient myths, but rather re-interpreted in light of its own 'not'.

But the sentence from Nancy I have just quoted tells us something more—something more precise—which we could summarize in the following way. The outcome of the extreme completion of nihilism—the absolute uprooting; the unfolding of technology; the integral character of globalisation—has two faces which we should not only distinguish, but also make interact: we could say that community is nothing else than the limit that separates and, at the same time, links them. On the one hand, sense appears to be lacerated, stretched out of shape, desertified—and this is the destructive aspect which we know so well: the end of any generality of sense, and the loss of mastery over the overall meaning of our experience. But, on the other hand, this very deactivation, this devastation, of general meaning opens the space of the contemporary world to the emergence of a singular sense that coincides precisely with the absence of sense and, at the same time, reverses it into its opposite. It is precisely when every given sense—located in a basic framework of reference—disappears that the sense of the world as such makes itself visible, reversed in its outside, with no reference to any sense, or meaning, that transcends it. Community is nothing other than the border, or transition, between this immense devastation of sense and the necessity that each singularity, each event, each fragment of existence must be in itself meaningful. It refers back to the character, both singular and plural, of an existence freed from any presupposed, or imposed, or postponed sense; of a world reduced to itself, able to be simply what it is: a

---

14. Jean-Luc Nancy, *The Sense of the World*, trans. Jeffrey S. Librett, Minneapolis, University of Minnesota Press, 1997, p. 62.

planetary world, without direction or cardinal points. A nothing-else-than-world. And it is this nothing in common which is the world that associates us in the condition of exposition to the hardest absence of sense and, at the same time, to the opening of a sense yet to be thought.

Translated by Lorenzo Chiesa

6 BEYOND NIHILISM: NOTES TOWARDS A CRITIQUE OF LEFT-HEIDEGGERIANISM IN ITALIAN PHILOSOPHY OF THE 1970s

Matteo Mandarini

> In the context of this seminar, the term 'metaphysics' indicates the tradition of thought that conceives of the self-founding of being as negative foundation. Whether or not an integrally and immediately positive metaphysics is possible (such as the one that ... A. Negri finds in Spinoza), remains an open question.
>
> Agamben, *Language and Death*

I

For metaphysics, the foundation is that upon which being rests, it is the foundation (*Grund*) that allows being to take place. But, 'as much as being takes place in the nonplace of the foundation (that is, in nothingness), being is the unfounded (*Das Grundlose*)'.[1]

Italian philosophy from the late 1960s to the 1980s—but this is by no means over—stitched a line leading from Schopenhauer and Nietzsche through to Wittgenstein and Heidegger that wove together *Das Grundlose* of being with the trajectory of nihilism. The very different theoretical and political backgrounds of the participants in these debates takes nothing away from the overall tendency to transfigure the foundation by stripping down

---

1. Giorgio Agamben, *Language and Death: The Place of Negativity*, trans. Karen E. Pinkus with Michael Hardt, Minneapolis, University of Minnesota Press, 1991, pp. xiii—translation modified. Also see Giorgio Agamben, *Il linguaggio e la morte. Un seminario sul luogo della negatività*, Turin, Einaudi, 1982, p. 6.

being and, ultimately, authorising philosophical mysticism and political opportunism.[2] The very real differences of the resulting positions—hermeneutical free-play, decentred community, or formalist decisionism[3]—cannot override the ultimate end of these tendencies: to provide a political (and rational) foundation for mysticism in terms of the immanent production of a merely residual, liminal negativity. It is these tendencies that I group together under the label of 'Left-Heideggerianism'—in recognition of their principal philosophical predecessor.[4]

What I propose to do here is to not detail the rich diversity of the theoretical trajectories of Left-Heideggerianism in Italian philosophy over the past thirty or so years—a daunting task and certainly not one to be attempted in a short article such as this. I intend, instead, to make some preliminary notes on something that has been largely overlooked in the discussion of recent Italian thought: i.e. the debate around the provocative assertions of Italian *Krisis*-thought. At the centre of this debate is Massimo Cacciari's *Krisis. Saggio sulla crisi del pensiero negativo da Nietzsche a Wittgenstein* published in 1976. That Cacciari's text was central to the development of a number of subsequent tendencies in Italian philosophy, political theory and political practice, is attested to by its influence on the develop-

---

2. By 'mysticism' I mean something very specific, as I hope will become apparent in the ensuing discussion.

3. I am tempted to include Agamben's image of the camp as the *nómos* of the modern and his notion of 'bare life'. The failure of Agamben's later project stems, as Negri argues convincingly in two recent essays, from the ontological indetermination, passivity and unproductivity of 'bare life' and not from a nihilistic foundation that he did so much to uncover in his earlier work. On this see: Agamben, *Language and Death: The Place of Negativity*, Giorgio Agamben, *Homo sacer. Il potere sovrano e la nuda vita*, Turin, Einaudi, 1995, Giorgio Agamben, *Homo Sacer: Sovereign Power and Bare Life*, trans. Daniel Heller-Roazen, Stanford, Stanford University Press, 1998, Antonio Negri, 'The Political Subject and Absolute Immanence', in Creston Davis, John Milbank and Slavoj Žižek (eds.), *Theology and the Political: The New Debate*, trans. Matteo Mandarini, Durham, Duke University Press, 2005, pp. xii, 476 p, Antonio Negri, 'Giorgio Agamben: the Discreet Taste of the Dialectic', in Matthew Calarco and Steven DeCaroli (eds.), *Sovereignty and Life: Essays on the Work of Giorgio Agamben*, trans. Matteo Mandarini, Stanford, Stanford University Press, 2006. Whether or not that saves Agamben from mysticism, i.e. from the re-establishment of a formal transcendence within immanence (and not in terms of a negative foundation that is the focus of this paper), remains, I would contend, 'an open question'.

4. I owe this label to Antonio Negri.

ment of 'weak thought' (*pensiero debole*) and, more importantly, on the notion of the 'autonomy of the political' as adopted by some of the leading intellectuals of the Italian Communist Party—amongst whom one must number Cacciari himself. Antonio Negri's critical review of this work in the Italian journal *aut aut*, which sparked the debate, did not conclude in any resolution or compromise between the contrasting positions. It did, however, serve to mark the point of irreducible conflict between two tendencies within Italian communist philosophy and politics. This debate cannot, then, be considered to be merely an incidental result of a review written for the Italian journal *aut aut* in 1976. Rather, it is fundamental to an understanding Italian philosophy and politics in a critical period of Italy's political and social history. It is also, something on which I shall focus in the second half of this paper, the point of convergence for a series of themes and problems that would be central to Negri's thought from that moment forth.

After discussing Cacciari's extraordinary book *Krisis*, I shall focus on a few selected texts of Negri's from the 1970s and early-to-mid 1980s. These challenge the specifically subtractive twist given to the Krisis of the foundation and they set the stage for Negri's continuing endeavour to develop a positive metaphysics which refuses *Das Grundlose* of Being (i.e. the determination of Being as negative foundation).

II

Massimo Cacciari, with whom Negri collaborated closely in the 1960s, was—along with Mario Tronti—instrumental in theorising the shift towards the 'autonomy of the political' as the political consequence of *Das Grundlose* of Being.

> Is it necessary, therefore, to make of Marxism the recovered philosophical foundation of science? But what does this foundation have to say to us today? Is it not, rather, a new dimension of *politics* that Marxism is able to open up for us—not in terms of a 'philosophy' of politics but as a 'will to power' exerted concretely over the multiplicity of languages of technology? Does one respond to Heidegger's and Nietzsche's thought through 'philosophy', appealing once again to Subjects, writing yet another meta-physical utopia for them? Or does one respond by starting to abandon the

rafts and ladders and penetrating, without emergency exits, into the politics of Technology, scientific research and into the infinite aporias of the 'social brain'?[5]

Cacciari's penetrating critique of the dialectic in the late 1960s and his analysis of a 'negative thought' that precludes any possible synthesis turned, in the 1970s, into an analysis of the means for the technocratic construction of 'new orders', founded on nothingness and crisis—a 'revolution from above' for the management of development by the representatives of the working class (i.e. through Italian Communist Party's control of the levers of political power). It is to this shift that we shall turn first.

III

> The form of the dialectic is the form of the negative that is affirmed positively—the recoverable contradiction. The *whole system* posits itself and maintains itself in terms [*nel segno*] of negativity: a movement of *universal* alienation is the true-real [*vera-reale*] totality.[6]

For Cacciari (and Negri) the Hegelian dialectic represents the highpoint in the victorious and expansive cycle of capitalist development, in which all contradictions, all conflicts are turned directly into productive moments of capital's advance as the self-realisation of Spirit. Everything becomes a moment of the production-consumption circuit of *Kapital-Geist*; the negative—in the form of 'determinate negation'—is the engine but it is an always already disciplined moment. That is, it is systemic and, hence, an integral moment—always presaging its disappearance—in the circuit of *Geist*.

In contrast to this 'virtuous' dialectic is the 'negative thought' developed in the nineteenth century by bourgeois theorists such as Schopenhauer, Nietzsche and Mach, and in the twentieth century by Wittgenstein and Heidegger amongst others. Cacci-

---

5. Cacciari's article in *Rinascita* quoted in Amedeo Vigorelli, 'Noi, i soggetti e il "politico". A proposito di Bisogni e teoria', *aut aut*, no. 155-156, 1976, pp. 196-203, pp. 196-7. *Rinascita* was the cultural and theoretical journal of the Italian Communist Party (PCI). See Massimo Cacciari, 'Noi, i soggetti', *Rinascita*, no. 27, 2 July 1976.

6. Massimo Cacciari, 'Sulla genesi del pensiero negativo', *Contropiano*, vol. 1, 1969, p. 131.

ari coins the term 'negative thought' in the late 1960s so as to precisely differentiate it from the positivisation of the negative that characterizes the dialectic. Negative thought begins by resisting all attempts by bourgeois ideology to pre-determine and synthesize. However, the role that the negative plays here is by no means straightforward. The negative is no longer immanent in the same way. It is no longer a moment of advance, no longer a dynamic moment produced and consumed at once. Cacciari claims that the process of alienation and recovery of *Kapital-Geist*, which makes itself at home in a world it produces, a world that it expels from itself only to re-appropriate more fully, is over. The negative now surrounds; it delimits and constrains but in so doing, it renders reality all the more 'ready-to-hand'. The Absolute Master, death now marks the outer perimeter of one's being and throws one back onto one's own-most possibilities, opening up an (instrumental) world for us and determining new orders to re-found the unfounded. The negative persists only in this paradoxical, marginal position that is the very condition of immanence but which, as we shall see, renders the mystical worldly— and the worldly mystical.[7]

Cacciari refuses to identify the mystical experience in the early Wittgenstein, for instance, with that of transcendence, on the ground that mysticism is—rather—the experience of (this side of) the limit (...of language, of my world...). It is the experience that the world is de-*limited*, hence, that it is 'radically worldly'.[8] He argues that the mystical should be opposed to the 'profound', on the basis that the profound is the un-sayable that one attempts to say—such as the attempt to provide a presentation (*Darstellung*) of the *noumenon*. Instead, the mystical stems

---

7. Mysticism is most obvious in the work of Schopenhauer and his denial of the Will, in Wittgenstein's *Tractatus* it appears explicitly, as itself, and it is there in Heidegger's 'guilt' and 'call of conscience'. Cacciari shows precisely how these examples are more than merely suggestive of the mystical and to what extent they actually develop a thinking of the limit as the definition of the mystical. The link I allude to between Hegel and Heidegger's conception of death is drawn from what is perhaps Agamben's most brilliant book, *Il linguaggio e la morte*.

8. Massimo Cacciari, *Krisis. Saggio sulla crisi del pensiero negativo da Nietzsche a Wittgenstein*, Milan, Feltrinelli, 1976, p. 95, Massimo Cacciari, *Dallo Steinhof. Prospettive viennesi del primo Novecento*, Milan, Adelphi, 1980, pp. 135-40, Massimo Cacciari, *Posthumous People: Vienna at the Turning Point*, trans. R. Friedman, Stanford, Stanford University Press, 1996, pp. 97-101.

from negative thought's excision of all reference to a 'real world' transcending the limits of language and that, thereby, enables a multiplicity of *technics* to internalize those very limits, autonomising themselves in a variety of specific 'language games'.⁹ They do so does so to make the world sayable, formulable and, so, to make it ready-to-hand as a function of a Will to Power. Negative thought is not, then, the attempt to express the inexpressible, to reach the unconditioned or absolute—an aim so prevalent in the history of metaphysics. Negative thought attacks all synthesis, all equilibrium and all reference: the real world becomes a fable, becomes ideological. The world's limits coincide with the limits of language, of what can be formulated. 'Our language games cannot be "situated" [*appaesabili*] ontologically'.¹⁰ Negri summarizes this as follows:

> Nietzsche's and Wittgenstein's work ... is reconceived in terms of a formal and negative thought but that is, thanks to the combination of the two elements, also constructive. It is constructive of logical and systemic horizons within which the efficacy of signification [*significativa*] is reduced entirely to the validity of the project, to the coherent rule of linguistic development [*alla regola coerente dello sviluppo linguistico*] and of the formal intention that constitutes it.¹¹

Thus, for Cacciari, the rational lacks all exogenous foundation. There is no *Ratio* to be sought in the world—all we have is a proliferation of rationalities, of 'language games', of ideological structures irreducible one to another, that are circumscribed by a nothingness. This is succinctly summarized by Giuseppe Cantarano's phrase, 'reason is nihilism inasmuch as it is the *historical project of the annihilation* [annientamento] *of being*'.¹² Hence

---

9. For Cacciari's discussion of the opposition between the mystical and the profound, see *Krisis. Saggio sulla crisi del pensiero negativo da Nietzsche a Wittgenstein*, p. 112.

10. Massimo Cacciari, '"Razionalità" e "Irrazionalità" nella critica del Politico in Deleuze e Foucault', *aut aut*, no. 161, 1977, pp. 119-33, p. 132.

11. Antonio Negri, *La macchina tempo. Rompicapi Liberazione Costituzione*, Milan, Feltrinelli, 1982, p. 41. This review essay was first published in the journal *aut aut*, nos. 155-156, 1976, with the title '*Simplex sigillum veri*. Per la discussione di *Krisis* e *Bisogni e teoria marxista*'. It was then reprinted in 1982, with the title 'Sul metodo della crisi filosofica', as chapter 2 of *La macchina tempo*.

12. Giuseppe Cantarano, *Immagini del nulla. La filosofia italiana contemporanea*, Milan, Bruno Mondatori, 1998, p. 319.

the indissoluble link between the 'mystical'—as described by Cacciari—and the mathematization, formalization of reality.

> This is how the essence of the 'mystical' appears [*suona*]. It is the simple description, which has been able to fully internalize its limits and that contains and shows [*mostra*] the nothing that embraces it, without saying any of it [*senza dirne un solo accento*].[13]

We can formulate and manipulate this 'conventional' world, precisely because the world is nothing but what can be formulated, beyond that is Nothing, which circumscribes and conditions without itself being said: '"behind" the different games *there is nothing*'.[14] Here Nietzsche, Wittgenstein and Heidegger are indissolubly linked: will to power, formalization of language and metaphysics as the reduction of Being to (formulated) beings—technology and power. This is summed up in one of Cacciari's most memorable and unsettling phrases: '[To have] power is to be integrated into the system'.[15]

Cacciari's political programme rests precisely on such a de-ontologized, even skeletal, grasp of actuality (*Wirklichkeit*). How else is the autonomy of the political to be understood if not as the decisionistic management of the multiplicity of fragmentary rationalities, as the working class—in the form of the PCI—taking control of the administration of the state, making up for a 'deficiency in rationalisation ... the inefficiency of the political apparatus'?[16] Political decisionism or, more precisely, voluntarist formalism situates itself in the place of the negative.

Also because we are not speaking of the autonomy of *a part of power* in relation to other parts; but of the autonomy of *all of*

---

13. Cacciari, *Krisis. Saggio sulla crisi del pensiero negativo da Nietzsche a Wittgenstein*, p. 112. Cacciari argues that, for all the differences that exist between the Wittgenstein of the *Tractatus* and that of the *Investigations*, it is precisely the notion of the mystical which opens the way for the development of the concept of language games, by effectively isolating the process of formalisation—reducing logical propositions to tautologies—and, thus, preventing the referent from acting as a unitary point of synthesis for the multiplicity of language games.

14. Massimo Cacciari, 'Critica della "autonomia" e problema del politico', in V.F. Ghisi (ed.), *Crisi del sapere e nuova razionalità*, Bari, De Donato, 1978, pp. 123-35, p. 131.

15. Cacciari, *Krisis. Saggio sulla crisi del pensiero negativo da Nietzsche a Wittgenstein*, p. 66.

16. Mario Tronti, *Sull'autonomia del politico*, Milano, Feltrinelli, 1977, p. 11.

power with respect to everything else that is not power; that is, to the rest of society. Hence, the autonomy of power with respect to what is or, better still, what was or was considered—generally—the *foundation* of power.[17]

In place of the foundation, then, we have the *beginning* of command over a process of rationalisation. The space of the Political is the space *between* language games, which negotiates their insoluble autonomy—which supports the negative that, in turn, determines their self-sufficiency.

> Let us, therefore, understand the autonomy of each technology, of each game, to mean that it possesses *only* one-law-of-its-own [*una-propria-legge*] (which is the result of an infinity of variations, which has been played and re-played, which is transformable and in-transformation because it is played). Let us understand the term 'autonomy' in this sense of *limit*.[18]

Paraphrasing Sergio Givone, it is only once one has abandoned faith in a political subject as foundation of revolutionary political change that one can rediscover a professional political class that can take over the administration of the actual to bring change from above:[19]

> The decision is preceded or pre-comprehended [*precompreso*] by nothing. Nothingness is the foundation of the decision.[20]

To what extent, then, does Cacciari succeed in escaping the metaphysical closure through this refusal of the ontological foundation produced by the saturation of rationalities in nothingness? This question is answered by Giorgio Agamben in his early book, *Language and Death*:

---

17. Tronti, *Sull'autonomia del politico*, p. 9.
18. Cacciari, 'Critica della "autonomia" e problema del politico', p. 130.
19. Technically Cacciari would be correct in refusing to see the activity of the PCI as coming 'from above' since his account in *Krisis* refuses any pre-determined hierarchy. On the other hand, by—along with Mario Tronti—viewing political power as something fundamentally independent from the 'rest of society' (Tronti, *Sull'autonomia del politico*) and arguing for the need for the PCI to garner that power *in order* to effect political change, the space for the autonomy of state-driven political processes is prepared. This argument is central to Negri's philosophical and political critique of the autonomy of the political and *Krisis*-thought.
20. Sergio Givone, *Storia del nulla*, Bari, Laterza, 1995, p. xxi. I would like to thank Alberto Toscano for bringing this important, although very problematic, book to my attention.

Today we live on that extreme fringe of metaphysics where it returns—as nihilism—to its own negative foundation (to its own *Ab-grund*, to its own unfoundedness). If casting the foundation into the abyss does not, however, reveal the *ethos*, the proper dwelling of humanity, but is limited to demonstrating the abyss of *Sig* [silence], then metaphysics has not been surpassed, but reigns in its most absolute form.[21]

For the structure that defines metaphysical reflection on Being (including Heidegger's—as Agamben shows so well), stems not so much from *foundationalism* as such, as from *self*-founding as *negative* foundation. For Heidegger, being destines but withdraws behind that which it destines. This withdrawal, the fact that being opens a clearing but recedes behind that which it clears, is analogous to the mystical as Cacciari describes it. We could argue that Cacciari repeats the *logic* of transcendence through the fabrication of a negative foundation (the mystical limit encircled by nothingness). It is no use his claiming that the mystical does not *found* the world but merely *delimits* it, that there is no receding being, for his actuality—the set of rationalities, of new rationalized orders—is nevertheless borne, supported by the nothing that surrounds the limits of the various language games in their very being formulated. In so doing the infinite movement of immanence is contained and constrained and we are left with the manipulation of dead terms by professional technicians of actuality.

Before unpacking the consequence of these manoeuvres it is important to explore this relation to Heidegger a little further.

IV

In 'What is Metaphysics?', and in the famous 1943 'Postscript', Heidegger does more than flirt with the identity of Being and nothingness:

> As that which is altogether other than all beings, being is that which is not. But this nothing essentially prevails as being ... we must prepare ourselves solely in readiness to experience in the nothing the pervasiveness of that which gives every

---

21. Agamben, *Language and Death: The Place of Negativity*, p. 53. Also see Agamben, *Il linguaggio e la morte. Un seminario sul luogo della negatività*, p. 67.

being the warrant to be. That is being itself.²²

Nihilation is not some fortuitous incident. Rather, as the repelling gesture toward beings as a whole in their slipping away, it manifests these beings in their full heretofore concealed strangeness as what is radically other—with respect to the nothing.²³

Heidegger refuses to render nothingness unthinkable, as has occurred in the metaphysical tradition since Parmenides, but he does so only by conjoining being and nothing. Whether he does this by presenting the 'occurrence of nihilation in the essence of Being itself'²⁴ or by claiming that 'nothingness appears to be the foundation of being'²⁵—in two rival formulations that are closer than at first appears given Givone's concession that the being 'that is preceded by nothing, that is determined by nothing ... at bottom [in fondo] is like nothing'²⁶—Heidegger is unable to fully satisfy Cacciari's demand that the nothing not be positivized. Cacciari's response is to forget being and Heidegger's nothing (das Nichts), and to affirm in a perverse appropriation of Heidegger—and yet against him—the 'actuality of the actual' (die Wirklichkeit des Wirklichen).²⁷ Cacciari's idea is to turn the negative, not into a positive element of language-games, but into their residual, produced condition. It is at once inactive, derived, and foundational. To fully pervert this appropriation, Cacciari affirms what has been best described by Heidegger as 'exact thinking' against the latter's demand for 'essential thinking':²⁸

All calculation lets what is countable be resolved into something that can then be used for subsequent counting. Calculation refuses to let anything appear except what is countable. Everything is only whatever it counts. What has been counted in each instance secures the continuity of

---

22. Martin Heidegger, 'Postscript to "What is Metaphysics?"', *Pathmarks*, ed. and trans. William McNeill, Cambridge, Cambridge University Press, 1998, p. 233.
23. Martin Heidegger, 'What is Metaphysics?', in William McNeill (ed.), *Pathmarks*, trans. David F. Krell, Cambridge, Cambridge University Press, 1998, p. 90.
24. Dennis J. Schmidt, *The Ubiquity of the Finite: Hegel, Heidegger, and the Entitlements of Philosophy*, Cambridge, MIT Press, 1988, p. 90.
25. Givone, *Storia del nulla*, p. 200.
26. Givone, *Storia del nulla*, p. 205.
27. Heidegger, 'Postscript to "What is Metaphysics?"', p. 231.
28. Heidegger, 'Postscript to "What is Metaphysics?"', p. 235-6.

counting. Such counting progressively consumes numbers and is itself continual self-consumption. The calculative process of resolving beings into what has been counted counts as an explanation of their being .... Calculative thinking compels itself into a compulsion to master everything on the basis of the consequential correctness of its procedure.[29]

One final lengthy quotation from Heidegger's 'Postscript' I hope will confirm my interpretation of Cacciari's peculiar faithfulness to Heidegger:

> Understood as a fundamental trait of the beingness of beings, 'will' is the equating of beings with the actual, in such a way that the actuality of the actual comes to power in the unconditional attainment of pervasive objectification .... As a way of objectifying beings in a calculative manner, modern science is a condition posited by the will to will itself, through which the will secures the dominance of its essence.[30]

Precisely the denial of a natural *Ratio*, of any structuring *Aufhebung*, means—for Cacciari—that there is no pre-given ought (*Sollen*), whether ethical or logical, by which irreducible heterogeneity can be reduced or can be reconciled once and for all, but only *Wille zur Macht* as the:

> ... *vital* necessity to com-prehend, order [*sistemare*], logicize the world, to *have power* over it .... Power is not synthesis—were it synthesis, there would no longer be any need for power.[31]

As Negri makes clear, what we are then left with is a calculable and manipulable set of elements, circumscribed by nothingness that delimits the serialized elements into language-games or rationalisation procedures, all of which are organized by a political decisionism—Will to Power, Will to Rationalisation—that determines the:

> ... historical necessity ... of a political class and a professional political class to which the management [*gestione*] of power is to be entrusted. ... In this way arises the moment of a war of manoeuvre [*guerra manovrata*], made-up of successive

---

29. Heidegger, 'Postscript to "What is Metaphysics?"', p. 235.
30. Heidegger, 'Postscript to "What is Metaphysics?"', p. 231.
31. Cacciari, *Krisis. Saggio sulla crisi del pensiero negativo da Nietzsche a Wittgenstein*, p. 65 & 9.

moves, all of which are scientifically calculated [previste] and tactically prepared.[32]

We can see, then, how Cacciari's re-conceptualisation of the notion of 'mysticism' serves an unsettling political project: to employ 'mysticism' for the task of a political *technics*—to the point of in-distinction of power and formal/ul-isation, and so to a technocracy of political action in which effectiveness is all. This will become the core focus of Negri's violent critique. Thus, Cacciari's thought shows a paradoxical adoption and disavowal of Heidegger. The withdrawal of Being, its retreat, 'ground[s] ... the dimension of being in its difference with respect to the entity'.[33] As we have seen, for Cacciari, this ontological difference results in the advent of language games that *confirm* that Being has always already only ever been understood in terms of beings (although Cacciari severs the etymological link between 'Being' and 'beings'). It is this that permits the reduction of politics to efficacy, to technology. In other words, we could say that the actual 'forgetting of Being' in Cacciari is both the condition of and conditioned by ontological difference, but for *Krisis*-thought this 'forgetting' frees itself of any sense of loss (or the possibility of recollection), such that the forgetting of Being—as the condition for beings to be, to be formulated and utilized—is, beyond this—and in contrast to Heidegger[34]—nothing, a nothing that circumscribes and (de-)limits, making possible, manipulable. It is as though Cacciari asks us to climb up and through Heidegger's propositions on the meaning and forgetting of Being, only to then 'throw away the ladder after he has climbed up it'.[35] What is left is the ontic world of the merely formulated, the calculable world of the will to will, which in being formulated, becomes utilisable.

The different language games co-exist but between them

---

32. Tronti, *Sull'autonomia del politico*, pp. 17-8.

33. Agamben, *Language and Death: The Place of Negativity*, p. 85. Also Agamben, *Il linguaggio e la morte. Un seminario sul luogo della negatività*, p. 105.

34. Agamben points out that Heidegger aims to think Being outside of its relation to beings (i.e. beyond metaphysics as he defines it) through the concept of 'Appropriation' (*Ereignis*) in his essay 'Time and Being' in Martin Heidegger, *On Time and Being*, trans. Joan Stambaugh, Chicago, University of Chicago Press, 1972.

35. Ludwig Wittgenstein, *Tractatus Logico-Philosophicus*, trans. D. F. Pears and B. F. McGuinness, London, Routledge & Kegan Paul, 1961, prop. 6.54.

there is no chance of synthesis, no possible ontological resolution but only the persistence of conflict and the need to negotiate it:

> Reality [*Reale*] is logicalisation [*logicizzazione*], Rationalisierung, which refuses the metaphysics of Language, the logic of *reductio ad unum*, the idea of the substance-subject—that takes on board the whole weight of the contradictoriness of the processes, of the multiplicity of languages, constituting its space and, so, allowing its form to emerge.[36]

On the one hand, the Political is a language game like the others, with its own specific rules and immanent possibilities of transformation and, on the other hand, it has other language games for its content. The Political then situates itself in such a way as to keep the confrontation between the various language games continuously open. It '*imposes* this continual confrontation, it prevents any game from withdrawing [*sottrarvisi*] from it'.[37] The only possibility is an endless compromise between different autonomies—between different language games characterized by the laws that specify them.

In *Krisis* we see the result of negative thought's refusal to give in to the temptation of dialectical resolution, to what Cacciari terms the 'recoverable contradiction'[38] that turns all antagonism into a moment of the development of the system of capital. The critique of the dialectic by Schopenhauer and Nietzsche that Cacciari had discussed in his important essay of 1969, enabled him to pin-point the *positivisation of the negative* as what was at stake in bourgeois thought. But what Cacciari was after in his 1976 book, through his analysis of Nietzsche, Wittgenstein, etc., was a way of re-conceiving the negative such that it would no longer be thought of as a moment by means of which the system develops itself and to turn it. Instead, it develops into a barrier that can be perpetually displaced and consumed as a moment of expansion of domination—of the Will to Power. Cacciari shows that all preceding notions of the negative end up neutering it, always already virtually resolving it—making conflict little more than an epiphenomenal form hiding a fundamentally pacific unity. This

---

36. Cacciari, *Krisis. Saggio sulla crisi del pensiero negativo da Nietzsche a Wittgenstein*, p. 185.
37. Cacciari, 'Critica della "autonomia" e problema del politico', p. 133.
38. Cacciari, 'Sulla genesi del pensiero negativo', p. 131.

view of the dialectic as—in its classical Hegelian form—fundamentally reactionary, is one that Negri largely shared with Cacciari.[39] However, Cacciari's solution merely served to confirm the theoretical and political break that had already taken place between them.[40]

V

> In this way, a cynical conventionalism—placed between an unstoppable [*irrefrenabile*] logicalising pressure [*pulsione*] and an hypocritical postulation of the mystical—represents the ruling class's gradual *prise de conscience* of the passage to capital's real subsumption of social labour and to negate the antagonism that sustains [*sostanza*] that passage as well as the claim [*rivendicazione*] to ontological truth the social subject expresses.[41]

The object of this scathing attack is Wittgenstein, but it is clear that the name 'Wittgenstein' also denotes Negri's erstwhile collaborator, the author of *Krisis*. One may, perhaps, free the power of the negative from *positivisation* by consigning it to the role of 'determining factor in the process of integration and

---

39. Negri's own relationship to the dialectic is extremely complex and cannot be easily summarized. His peremptory tone when discussing it is quite often misleading, as is the all too hasty suggestion that Negri refuses the dialectic in a manner analogous to Deleuze and Foucault. This is both false and—ultimately—fails to shed light on any of these thinkers' take on the question. I have discussed Negri's nuanced conception of the dialectic in Mandarini 2005.

40. For all the criticisms Negri would direct at *Krisis*-thought, he recognizes the 'wonderful attempt to positively recuperate the efficacy of negative thought' as late as 1981 (see Antonio Negri, *The Savage Anomaly: The Power of Spinoza's Metaphysics and Politics*, trans. Michael Hardt, Minneapolis, University of Minnesota Press, 1991, p. 211ff. & n.3). This is well after the vigorous critique directed at *Krisis* in *aut aut* and of his angry tirades against 'Nietzsche in parliament' (see Antonio Negri, 'Domination and Sabotage', in Timothy S. Murphy (ed.), *Books for Burning: Between Civil War and Democracy in 1970s Italy*, trans. E. Emery, New York, Verso, 2005), which followed the election of Cacciari to the Italian parliament in 1976 under Berlinguer's strategy of 'historic compromise' between the Italian Communist Party and the ruling Christian Democrats. It is clear that Negri is affirming negative thought's refusal of dialectical synthesis, of domestication or positivisation of the negative. But to stop there, he will argue, is to remain within the formal antinomies of thought and to subordinate practice to technocratic negotiation or national compromise.

41. Negri, *La macchina tempo. Rompicapi Liberazione Costituzione*, p. 33.

rationalisation'.[42] But is not the result of this that the negative becomes domesticated? The process of de-ontologisation, that is, of the excision of the referent that allows the multiplicity of formal, conventional rules to be deployed, as a pure free-floating *technics* of manipulation and efficacy, reduces thought to what works and, hence, to the apologetic subordination to existing states of affairs or—at best—to a 'fetishistic overdetermination'[43] from above, i.e. to ideology and political opportunism. Once one excises all ontological foundation, power is necessarily defined by the level of integration into the system, by one's ability to 'work it'.

> It is not the degree to which one approaches an illusory substance but the degree of integration with which it operates in the process of rationalisation [that] decides the value and the power of logical form.[44]

According to Cacciari, substance is illusory, Being is equally so—both represent merely utopian moments of synthesis. In their place there is nothing. Nothing circumscribes and conditions the wholly immanent nature of the conventional, formal rules—thus establishing the worldliness of the mystical. But, as we have already seen, it is clear that this negative foundation, the condition for the 'concrete search for re-foundation', does not signal an escape from metaphysics or even from a constraining of immanent processes of change. Indeed, that the 'processes of re-foundation', of formalisation and conventionalisation are constituted as 'movements *internal* to the "negative"',[45] is by no means evident since the process of formalisation *presupposes* a negative foundation as denial of Being, Substance, etc. Thus the mystical, the Nothing that circumscribes, marks the formal condition for immanence but also *delimits* the immanent and, in so doing, turns the negative into the presupposed product of the very process it must condition. The 'movements internal to the "negative"' may be immanent but the negative remains abstract, unrelated, undetermined and a merely manipulable epiphenomenon.

---

42. Cacciari, *Krisis. Saggio sulla crisi del pensiero negativo da Nietzsche a Wittgenstein*, p. 8.

43. Negri, *La macchina tempo. Rompicapi Liberazione Costituzione*, p. 43.

44. Cacciari, *Krisis. Saggio sulla crisi del pensiero negativo da Nietzsche a Wittgenstein*, p. 68.

45. Cacciari, *Krisis. Saggio sulla crisi del pensiero negativo da Nietzsche a Wittgenstein*, p. 8, my emphasis.

In short, Negri suggests that Cacciari pays a heavy price for having saved the negative from its positivisation in the development of Capital-*Geist*—he effectively domesticates it. He has been able to maintain the insolubility of crisis and prevent any easy synthesis, but—as Negri points out in his 1976 review—he has done so while losing any concrete conception of the negative, losing the ability to analyse struggles and ending up with a fundamentally domesticated, opportunistic conception of the negative and of politics. The problem for this epigone of the autonomy of the political, is that the moment of decision and the subject of decision cannot be understood independently of the process of rationalisation. Givone argues that in founding being on nothingness and thereby allowing beings to appear in their difference from being, as 'not being nothing',[46] Heidegger thereby establishes the possibility of freedom:

> ...precisely because to be 'immersed in nothing', Dasein is always already beyond the entity, beyond the world .... To be immersed in nothing means to transcend ... transcendence is freedom.[47]

But what happens if the nothing ceases to found being and instead becomes merely a manipulable element to be deployed, or a residual effect of rationalisation procedures? This effectively collapses the problem of the relation between the autonomy of the antagonistic class subject into that of its organization, since the subject is defined merely by its ability to effectively negotiate the formal rules of the multiplicity of languages and so cannot be an object of analysis independently of those formal rules. As Negri argues in his review, the problem of the relation between class autonomy and its political organization is not thereby resolved but merely exorcized by transferring autonomy to the ideological structures or language games/conventions and the level of its organization is defined precisely by the effectiveness of those same formal structures. The truth of a language game or rationalisation procedure is given by the principle of efficacy that is determined by the level of organization of the language game ... i.e., by its efficacy. In Negri's words:

> The complete sophism is: the guarantee of truth of

---

46. Givone, *Storia del nulla*, p. 199.
47. Givone, *Storia del nulla*, p. 200.

organization is given by the principle of reality that only that organization can guarantee.[48]

Autonomy collapses into organization and organization into effective management. For Cacciari, whether the working class or the capitalist class gains power is merely a question of efficacy, of degrees of integration.

VII

What are we then to make of the usage of the negative in Negri's thought? In the small space that I have left, I can only hope to outline the skeleton of an alternative that I believe can be uncovered in Negri's writings. I shall attempt to summarize this in some baldly stated theses:

The question of the nature and position of the negative is *the question of politics*: specific struggles between classes determine the nature and position of the negative. Conversely, the question of the nature and position of the negative has concrete political effects, i.e. co-determines particular relations between classes in struggle. How the negative is played out in struggles between classes—in the form of antagonism, contradiction, terror, or alternation[49]—is, therefore, intimately linked to the question of politics.

We have already seen that the question of foundation at issue in metaphysics *cannot be understood independently of the nature and position of the negative*. For the question of foundation—so crucial to the history of metaphysics—is intimately related to that of the position of the negative.[50]

For Negri, the question of metaphysics cannot be grasped independently of the question of politics—*and vice versa*. Moreo-

---

48. Negri, *La macchina tempo. Rompicapi Liberazione Costituzione*, p. 48.

49. One fundamental contribution to the political function of the negative has been provided by Mao, for whom the negative, or antagonism had to be comprehended in a complex interplay of principle and secondary 'contradictions'. See for example the analysis in Mao Tse-Tung, *On the Correct Handling of Contradictions Among the People*, Peking, Foreign Languages Press, 1957.

50. For, in the history of metaphysics, being is always conceived in terms of the question of foundation, even when that foundation is entirely negative. In the words of Giuseppe Cantarano, 'the nothing has always supported the stability of being. The nothing [*Das Grundlose* of being] is the foundation of being'. See Cantarano, *Immagini del nulla. La filosofia italiana contemporanea*, p. 305.

ver, both the question of metaphysics and that of politics are intimately related to the question of the negative.

As a corollary to this: *the question of the nature and position of the negative is the question of method.* Where 'method' is understood as immanent to the real, as a practice that is ontologically constitutive—*politics as metaphysics as ontology*: the *'real* movement which abolishes the present state of things'.[51] This is what Negri means when he speaks of a method that

> ...dispenses with all that remains of the exterior, gnoseological, and methodical connotations in order to become a substantial element, a constitutive key to the world. If this is a method, it is the method of being.[52]

For such a method involves situating the negative *within* the specific antagonism of class forces within a determinate, i.e. a concrete, social formation—and projecting the specificity of that antagonism, i.e. of the nature and position of the negative, into alternative standpoints of metaphysics and of politics. That is:

> When capital constitutes the political as the domination of one class by another, metaphysics is affected [*subisce*] by both poles of the relationship: it is the forces in struggle that assume the sense of a metaphysical tradition and oppose it to another one.... A metaphysics, distinct metaphysical positions and the alternatives they represent are the most concrete of historical objects. They are 'concrete' because they are swollen [*gonfio*] with antagonisms and possibilities.[53]

The history of the transformations of the nature and the position of the negative is the history of the antagonism between 'blessed' versus 'damned' metaphysics of which Negri speaks in his Spinoza book. Thus, ultimately, it is the reflection of class struggle.

---

51. Karl Marx and Friedrich Engels, *The German Ideology*, Moscow, Progress Publishers, 1976.

52. Negri, *The Savage Anomaly: The Power of Spinoza's Metaphysics and Politics*, p. 150. Also see Antonio Negri, *L'anomalia selvaggia*, Milan, Feltrinelli, 1981, p. 182.

53. Antonio Negri, 'Note sulla storia del politico in Tronti', *L'anomalia selvaggia*, Milan, Feltrinelli, 1981, p. 290. The essay 'Note sulla storia del politico in Tronti' was published as an appendix to Negri's *L'anomalia selvaggia* along with another two short articles.

VIII

Negri argues that the only way to concretize the negative, to not leave it entirely in the hands of the theorists of the mystical, is to conceive the negative as *immanent* to struggle, i.e. in terms of the specific characterisation of the negative within class struggle. But by so doing, does he not end up returning the negative to its subjection to the dialectical *Aufhebungen* and so to the development of *Kapital-Geist* or—at best—*Kommunismus-Geist*? For is the negative able to escape synthesis, i.e. does it not merely get *resolved* one way or another in the result (as is the case with determinate negation)? In either case, is not the result a final pacifying *telos* with all that it entails? That is, does not the negative become literally nothing, i.e. it is absent for it is always already accounted for, reduced, *aufgehoben* in the result, thus effectively repeating Parmenides' inaugural gesture of the state tradition of metaphysics, whereby the source of all conflict is to be excised to leave us with the One?

> Being is ungenerated and imperishable, entire, unique, unmoved and perfect; it never was nor will be, since it is now all together, one, indivisible.[54]

This, the 'blessed' tradition of state-thought, is the ancient but still active origin of bourgeois thought.[55] *Contra* Cacciari then, bourgeois thought is, rather, defined as one where the horizon of war is perpetually refused in favour of security, where the negative is excluded from the commonwealth, indeed where the commonwealth is entirely *constituted* by a foundational exclusion of the horizon of war. This is the fundamental problem of reactionary thought (Hobbes *per tutti*). Instead, Negri wants to champion that other thread (of politics and of metaphysics), which views

> ...war as the fundamental and insuperable condition: where

---

54. Parmenides and A. H. Coxon, *The Fragments of Parmenides: A Critical Text with Introduction, and Translation, the Ancient Testimonia and a Commentary*, trans. A. H. Coxon, Assen, Van Gorcum, 1986, pp. 60-2, frag. 5.

55. Is this *still active origin* not evident even in that danger against which Tronti cautions us, even as he proposes his notion of the autonomy of the political, as the 'risk of a more organic relation [*azione*] between the state and capital, the danger of a formidable power-block that—at that point—could not be attacked and would be invincible'? Quotation from Tronti, *Sull'autonomia del politico*, p. 19.

it is not a case of eliminating it but of making it function without precipitating into a simple massacre. Instead, making it operate against the relations of production and in favour of the productive forces and their free expansion.... Only by going back over the history of metaphysics, only by discriminating within it real alternatives do we have the possibility of contributing to the construction of new models for the refounding of class politics within antagonism.[56]

Once again, we see that the nature and position of the negative is the question of politics. For how are we to conceive of this 'within antagonism', i.e. how are we to comprehend the nature and position of the negative in a way that leaves it open, without resolution but without rendering it merely formal—without throwing us onto the mystical, and so opportunism, or back into the arms of the dialectic? This problem, I believe, is one that haunts Negri's thought for over four decades—from his writings on labour and the constitution, to his detailed work on the state-form, from his reappraisal of Spinoza through to his most recent reflections on time and ontology.

IX

Negri gropes his way towards a solution to this apparently extremely abstract (i.e. theoretical) but—as we have seen—completely concrete (i.e. political) problem in the late 1970s and early-to-mid 1980s. At this time, he argues that it is only by making the negative into an element of concrete practice and, therefore, ontologically substantial, that it can escape formalisation or auto-dissolution in a pacifying synthesis. The answer is not that negative thought must be rejected but that alone it is insufficient. In negative thought the negative is purely logical or ideological, i.e. it is parasitic upon that which it negates or, more precisely, its evacuation of all ontological foundation from what it critiques nevertheless enables the object of its critique to persist as the de-substantialized, de-ontologized form of languages and rationalities. Since none of them are invented, they can at best be re-articulated. Such an ideological negation is ideological in the strong sense: all that remains is ideology. Thus, Cacciari's negation allows the proliferation of ideologies as it removes their

---

56. Negri, 'Note sulla storia del politico in Tronti', pp. 291-2.

material support. With the excision of the ontological referent, ideological struggle becomes entirely formalistic, opportunistic and divorced from the subjects of struggle. Against Cacciari's intentions (but not so far away from his recent practice), this appears to be an early anticipation of the political logic played out today in the contemporary discourses of 'beyond left and right' and 'modernisation'—where to modernize is little more than to *make adequate* to the dominant conditions of accumulation and exploitation, while negating those conditions (ideologically). The Fordist factory—at least in the West—hardly exists anymore. Thus, as the current champions of the 'beyond left and right' argue, the referent, the space of exploitation as well as its subject, no longer exist. Exploitation is no longer a battleground, the battle today becomes the purely social one against 'social exclusion'. Poverty is thus ascribed to individuals' disconnection to a supposed space of possibility, of opportunity—an eminently ideological space from which the substantial ontological body of the exploited is excluded. What is demanded by the 'modernizers' is that the excluded be increasingly integrated into this rich space of possibility. The excluded must be able to learn and speak the different languages: 'It is *inexorable [inesorabile]* to learn to play a language if we want to experiment with its gaps, differences, limits and aporias ...'.[57] The specular double of Thatcher's 'there is no such thing as society' is the Blairite and Communitarian claim that all that exists is society. The question is how individuals can be made to participate *more* fully, *more* inclusively.

Is this so far from Cacciari's claim, '[To have] power is to be integrated into the system'?[58]

It is this logic that Negri defies by ontologising the negative. Subordinating philosophy and practice to Krisis, to *Das Grundlose* of the foundation, fails to pit the negative against Power—to generate any antagonism that cannot be compromised by it—and so it remains prisoner to Power. As Negri argues vociferously in various places in the mid-1970s, Krisis cannot be made to operate as motor or condition for a communist politics—for such a politics, this conception of the negative would forever sub-

---

57. Cacciari, 'Critica della "autonomia" e problema del politico', p. 127.
58. Cacciari, *Krisis. Saggio sulla crisi del pensiero negativo da Nietzsche a Wittgenstein*, p. 66.

ordinate it to the transcendence of Power. The 'historic compromise' proved an historic failure, as became increasingly evident in the course of the 1970s. The policy of the PCI became increasingly subordinate to that of the Christian Democrats (DC) to the point that the DC increasingly excluded the PCI from the levers of power while drawing the PCI into the fierce repression and criminalisation of a large number of the extra-parliamentary left. Subordination to the State became total. Yes, the conflict was not resolved in a pacifying dialectical synthesis, but the un-synthesisable discourses of the PCI and DC became elements of a *dispositif* subordinated to the maintenance of a means to maintain the continuity of dialogue, i.e. of this 'continuous confrontation'.[59] *Entrismo* quickly became *trasformismo*.[60] I seriously doubt Negri knew just how prescient his critique of Cacciari in 1976 would be.

Nevertheless, the years spent in prison following the infamous 'April 7th' verdict, were an extraordinarily fertile period theoretically for Negri and his endeavour to achieve a thought and practice of the negative that would integrate the lessons of negative thought while refusing the logic of integration and the correlative state-terrorist repression. It is interesting to see Negri take up again his study of seventeenth-century philosophy after a decade. To his previous study of Descartes (1970), Negri now adds his influential study of Spinoza. This may appear a strange way to approach the very pressing failure of communist (reformist and revolutionary) politics of the 1970s but Negri emphasizes the timeliness of this study by titling one of the final sections of the book, 'Negative Thought and Constitutive Thought'.[61] There he argues that, in contrast to Descartes, Spinoza refused to be satisfied with subordinating thought to its crisis and to de-ontologize the negative, and was able instead to give the negative its autonomy by turning it into an element of his ontology. So, against the 'reformist' strategy of Descartes, Negri postulates the constitutive and productive one of Spinoza.[62] Negri's strat-

---

59. Cacciari, 'Critica della "autonomia" e problema del politico', p. 133.

60. The PCI's attempt to find a point of entry into government through compromise with the DC resulted in the transformation of its policy into one of defence of governmentality.

61. See chapter 9, § 1 of Negri, *L'anomalia selvaggia*; Negri, *The Savage Anomaly: The Power of Spinoza's Metaphysics and Politics*.

62. See Antonio Negri, *Descartes politico. Della ragionevole ideologia*, Milan,

egy, then, is to suture negative to constitutive thought. It is to re-ontologize Krisis:

> If dialectics cannot be conceived as the form by which determination is resolved,[63] if—nevertheless—the terms of a dialectical problematic remain, and finite elements oppose one another without encountering *Aufhebungen*, what shifts, passages, relations will the existent terms have to experience [*che pure i termini dell'esistenza debbono conoscere*] on the negative edge of this situation? Certainly, it is not a case of a logical sequence; there is no linearity given on this horizon of being. In contrast, we encounter ruptures, crises, and suffering. *But all of this is given within being, against an ontological backdrop that contains and relates these emergent elements* [*emergenze*].[64]

It is clear, then, even in 1984-85 when Negri was completing his little-known but hugely significant book on Leopardi, that he was still trying to find a way to insert crisis and negation into ontology and so refuse the logic of *Das Grundlose* of being. Politically, the failure represented by Cacciari's *Krisis* was evident, and the 1980s and 1990s would only confirm the neutering of the negative once it is subordinated to the Political in terms of a 'continuous confrontation'. Theoretically, however, the problem remained.

Whilst accepting the rejection of the foundation that characterized negative thought, Negri would refuse to either neuter the negative or allow it to be resolved ideologically through the 'autonomy of the political'—i.e. by de-ontologising it. Instead, he would endeavour to turn it into an element for the production of new being:

Feltrinelli, 1970, Antonio Negri, *The Political Descartes: Reason, Ideology, and the Bourgeois Project*, trans. Matteo Mandarini and Alberto Toscano, London, Verso, 2007. In particular the 'Postface to the English edition of *The Political Descartes*'.

63. For Hegel determination is, of course, negation—and vice versa, since in the terms of his dialectic, negation stops being abstract and formal because it is always a determinate negation. This must be borne in mind to understand the full import of refusing a dialectical resolution of the determinate. Negri wants to maintain the concreteness of Hegelian negation while refusing its insertion within the neutralising logic of dialectical synthesis—a difficult balance to maintain.

64. Antonio Negri, *Lenta ginestra. Saggio su Leopardi*, Milano, Mimesis Eterotopia, 2001 [1987], pp. 44, my emphasis.

...reality as origin [as *arch*][65] is negated and it presents itself instead as a creative surface. ... There is only the revelation of the polarity of being and poetic subjectivity, within a relationship that negates every pre-existing reality and returns being to us as revolution, as radical transformation.[66]

In this way, 'Ontology becomes the science of the rupturing [*rottura*] of being'[67]—ontology is the science of revolution; revolution is the practice of ontology. We could also say, therefore, that ontology becomes the science of the negative. Thus, Krisis understood as de-ontologisation of the negative is refused—nothing, negation is instead understood as the potentiality of being[68] insofar as it refuses characterisation as stasis, constant capital, death.

Although Negri cannot be said to have resolved these difficulties to his satisfaction or ours, his work has been crucial in bringing to light this pivotal problem for the development of communist philosophy and politics. He also set out the markers that separate his own endeavour from those who have attempted to trace back to Heidegger the theoretical tools towards, if not a revolutionary, at least to a progressive politics. How pressing this problem remains for Negri and for any of us who wish to be able to think the political, to think the negative, is summarized in a recent preface Negri wrote to a book on Deleuze by Francesco Lesce:

> I believe that once all dialectical mediation is set aside, once Heidegger's hypostasis of being has been criticized, the problem of the negative reappears. How can one confront it *inside*, within, in the heart of materialist ontology? ... The negative is consistent [*consiste*]. How can it be assumed, resolved, how can one suffer it and destroy it in a world without an outside? How can the painful consciousness of the negative be grasped within and against the positive

---

65. When speaking of the origins of the Greek word '*arch* ', Reiner Schürmann reminds us that 'Aristotle is the one who explicitly joins the more ancient sense of *inception* with that of *domination*'. See Reiner Schürmann, *Heidegger on Being and Acting: From Principles to Anarchy*, trans. Christine-Marie Gros, Bloomington, Indiana University Press, 1987, p. 97.

66. Negri, *Lenta ginestra. Saggio su Leopardi*, p. 154.

67. Negri, *Lenta ginestra. Saggio su Leopardi*, p. 167.

68. 'There is no being other than the being that we produce'. See Negri, *Lenta ginestra. Saggio su Leopardi*, p. 215.

reconciliations of being?[69]

These intractable questions continue to assail contemporary thought and demand theoretical, and—more importantly—practical resolution.

---

69. Antonio Negri, 'Prefazione', in Francesco Lesce (ed.), *Un'ontologia materialista. Gilles Deleuze e il XXI secolo*, Milan, Mimesis, 2004, p. 6.

7 THE SYMBOLIC INDEPENDENCE FROM POWER

Luisa Muraro

In one way or another, anyone who approaches philosophy always has to deal with the problem of what it means to think the unthinkable, to dwell on that extreme and undefended edge where thought loses itself, faints [*viene meno a sé, s-viene*], something that is often compared to or even equated with death. It is the opposite of a border, which we experience in a pleasing manner as we drift off to sleep; rather, it is the experience of a thinking precariously balanced between collapse and delirium. This experience is also lived and reflected upon by people who are not philosophers by vocation, in mystical or artistic research, to mention familiar examples.

I want to consider a situation that is close to this, but also significantly different, one which is acknowledged and investigated in the context of the psychology of discovery, and which to my mind is of far broader interest. I have formulated it in the following terms: what happens to thought when it encounters the unthought? I use the past tense because I am taking up again an already existing line of inquiry.[1] In the original version, this was

---

1. In the first part of the article I take up again the inquiry that makes up Chapters 4 and 5 of my book *Al mercato della felicità. La forza irrinunciabile del desiderio [At the Market of Happiness: The Unrenounceable Force of Desire]*, Milan, Mondadori, forthcoming 2009. This text freely reproduces my talk at the colloquium of the group Diotima at the University of Verona, 10 October 2008. The general theme of the colloquium was 'Power and Politics are Not the Same Thing'. The presentation of October 10 in turn reprised a contribution of mine to the journal *Via Dogana*, 86 ('Libreria delle Donne di Milano', *Via Dogana*, no. 86, 2008), for an issue entitled *Il miraggio del potere nel de-*

my formulation: you can be deaf and nonetheless hear sounds thanks to the vibrations of the acoustic medium, you can be blind and intuit colours thanks to the magic of words; I wonder how thought can stop in its tracks and notice the unthought that is happening to it.

I proceeded by examining some texts that have nothing in common with one another, save the point we're concerned with, namely that they allow us to broach the situation of a thought that comes up against an unthought.

One is the story of a nurse, addressed to me in the context of a university course on the thought of sexual difference. Some years before, when she was still in training, helping an old patient to bathe, she ended up seeing for the first time, and with no prior warning, the genitalia of a hermaphrodite (a term she will get from an older nurse). The author of this story noted that during the bath, unusually, the wife of the patient was also present, 'as though she were keeping watch': but keeping watch over what? In my view, on the sexual identity of her husband, endangered by the gaze of the nurse on his genitalia, so much so that the nurse, in her story, does not call him a man but a 'human creature', eliciting my criticism.

The second text is the first act of Shakespeare's *Macbeth*: Macbeth, returning victorious from the battlefield where he has risked his life for his king, encounters three witches who inspire in him the idea of taking over the king's place. A desire which shakes him and the symbolic order to its foundations. With the wisdom of hindsight, which is to say with the mediations that later emerge, we recognize in that desire a mute anticipation of what will turn out to be the principal characteristic of modern democracy, according to which everyone, male and female (in abstract terms, of course), can aspire to any public position or status.

The third text was one I composed on the basis of some passages in Freud's letters and essays, and it shows the path that took him, during the period of his first hysteric patients, to deceive and undeceive himself regarding the *trauma* of childhood sexual seduction, in the process acquiring an ear for the unconscious. It is worth noting that *trauma* is a term that we can apply to all the events of the type considered here.

Finally, the fourth text is the tale of Paul's so-called conver-

---

serto della politica [The Mirage of Power in the Desert of Politics].

sion, as drawn from the Acts of the Apostles and his Epistles, and it reverberates in the reflection of Western Christian thought to this very day. I have taken into consideration recent texts linked to the discussion on so-called political theology and I have come to the conclusion that the Pauline trauma is once again making its effects felt.

I place myself among those who attribute a political thought to Paul, but in a sense that subverts the very idea of politics. More precisely, for me his thinking culminates in the idea of symbolic independence from power.

I have been led to this reading by the reflection on the relationship between politics and power in the women's movement and feminist thought. Despite all the confusion around State feminism, entirely aimed at fighting discriminations and instituting an equality between the sexes (which to my mind is practically impossible and perhaps senseless), the feminist movement revealed that the aversion for politics understood as competition and struggle for power, an aversion widespread among women, is not a refusal of politics, but on the contrary a demand for politics: there is a demand that where the machine of power now stands, political life should come to be.

In my inquiry into the unthought, the question that lay in the background was and remains the following: what happens to thought when the thinking subject is a woman, when it becomes aware that it is the thought of a female thinker [*una pensante*], that is to say a thought linked to being a body? What does this mean for thinking itself? Is it inconsequential or are there repercussions, and if there are, how do they manifest themselves in the order of the true/false, in the linguistic-expressive order (for instance, the 'I' that assumes/does not assume female gender predicates), in the pragmatic order (that is, the symbolic and practical efficacy of words)? And what becomes of our ineluctable *historicity*? With this term, which is not a synonym for historicism or relativism, I mean that what presents itself to us in our experience is never something that is absolutely self-identical, nor incontrovertible; pure thought thinks necessity, but experience does not experience necessity, so that thought is called to the work of mediation in order that what there is does not come to nothing. I would even say that this call constitutes the very essence of thought and, at the same time, its kinship with politics.

That the thinking subject is of female gender simply makes manifest the historicity of thought, in very precise and inexorable terms, which remind us of our being, all, men and women, born of woman. Thought is presented with its dependence on being a body [*essere corpo*], in the most difficult form, that of an insurmountable asymmetry: that women are born from woman and men instead ... also.

Whence a troubling of thought that feminist research, both historical and philosophical, has amply registered. Thanks to it, we know a lot of things about the trauma represented for the life of thought when it discovers itself as the thought of subjects called women on account of sexual difference, in other words, the discovery that women think and that what a woman thinks is thinking for all.

In effect, this research has given rise, in past and present culture, to a panoply of defensive reactions. There has been discredit and ridicule (think of Molière ridiculing women of culture) and insecurity (the specialist of women's biographies Carolyn G. Heilbrun has spoken in this regard of a 'rhetoric of uncertainty'). Many feminines have become masculines. Many contributions of women to human culture have been salvaged by becoming contributions of men—whether by feint, theft or error—since the male sex has historically committed itself to safeguarding the thinking transcendence from its being a body, with everything that follows in terms of objectivity, impartiality, universality.

What's more, feminism has contributed to showing how the 'safeguard' offered by the unique masculine, of an objective and universal thought, bears so many affinities with the patriarchal order, that is with a system of domination that has constituted itself into a veritable civilisation.

It should be said that feminism, or rather part of it, in its turn gave rise to new defensive reactions, promoting a view of the world in which women share power with men in a regime of perfect equality and heated competition. In American cinema there are plenty of female cops that surpass their male colleagues in terms of homicidal determination. The mass media, political parties, and intellectuals are pretty much aware only of this feminism, though it is minoritarian among women and remote from the beginnings of the feminist movement, in the sixties and seventies of the last century: just think of Carla Lonzi.[2]

The confusion between politics and power has today become extreme, putting politics into a state of agony. We should realize that politics is not something co-terminous with society: political life is an additional possibility of life in common and is given under certain conditions, so much so that there have been and can be societies without politics, where there are only power relations, contrast, envy and the desire for power, and where the best that humanity can conceive and realize is shunted to the side. We are moving in this direction, on an inclined plane that cannot be reduced to a question of democracy: faced with this drift, limiting oneself to the defence of democracy is a mistake which is making more than a few people of good will waste their time.

The thought that believes it can withdraw from historicity, which is to say from the relation from what is other than itself, the thought that refuses the relativity of relation, is a thought that can hide its fallaciousness only through the possession of a power over other human beings. This possession in effect exempts one from the work of necessary mediation. But it has its price, because the exemption from the work of necessary mediation is the premise of the typical stupidity that manifests itself in the wielders of power over others.

In the world there is, there has always been, a lot of feminine intelligence, nourished by the necessity of mediation, active in human traffics and brilliant in the exchanges with nature, a non-linear intelligence, full of ruses and expedients, closely related to cunning. But everything would have remained in the domain of servility without the feminist consciousness-raising which interrupted the continuity of a social order which offered emancipation to women as a goal to be attained.

With feminist consciousness-raising, the unthought ceased to provoke defensive reactions, in order to become food for thought, that which makes one think. The end of the confusion between politics and power is an outcome of this break or displacement that I like to call a 'dodge', to echo the title of the film *L'esquive* by Kechiche (France 2003). The awareness of being

---

2. [Carla Lonzi (1931-1982) was a feminist writer, pioneer of a feminism of self-consciousness and sexual difference. She was a founder of the group and publishing house *Rivolta Femminile* and the author of, among others, *Sputiamo su Hegel* [Let's Spit on Hegel] (1970) and *La donna clitoridea e la donna vaginale* [The Clitoral Woman and the Vaginal Woman] (1971).]

elsewhere and otherwise—this is what it means to be conscious of oneself: not letting oneself be found within the trajectories of power, within its predictions, exposed to its manipulations; to exist in relation and in the verbal exchange of an autonomous and liberating practice.

When thought, struck by the trauma of being the thinking of a *she*, does not react by defending itself in the ways that we're accustomed to, then trauma becomes a new beginning for oneself and a free sense of sexual difference. Our eyes and feelings are transformed: it is not simply a form of reasoning, but an experience. In light of this mutation, which is above all a political practice, the confusion between politics and power becomes visible, and the demand that this confusion be brought to an end arises spontaneously. We thus come to conceive of politics itself as a matter of gaining a free existence in spite of power. No: against power, but behind its back and to its detriment.

The free sense of sexual difference is like the agent of a distilling of politics; it shows how much will to domination there is in the aspiration to the universal, how much violence there is in the processes of objectivation, how much imbalance of power in the power of interpretation.

My main thesis, in this text and in this phase of my thinking, is that the radical antidote against confusion, for those who love politics and are trying to extricate it from power, is constituted by the symbolic independence from power, as we will now see.

The trauma of an unthought presenting itself gives rise to a number of manifestations, among which is the fear of deceiving oneself, which Shakespeare translates very effectively in Banquo's question.

'What, can the devil speak true?', Macbeth's comrade in arms asks himself, when the witches, prophecy begins to realize itself. He asks himself out loud, almost as if to get Macbeth to talk about the meeting with the witches, having noticed the profound turmoil that is affecting his superior.

Faced with the unthought, what an alert thinking fears is precisely deception. The danger is not that of being deceived, even though this too is present—because the deceivers who make you believe one thing rather than another do exist—but to deceive oneself because of something that is undeniably true. This is what the diabolical use of truth consists in, something that the inquisitors of all the world's churches specialize in. This is not the case with Macbeth, who is instead deceived by the 'truth' of his violent and sudden desire which he seeks to realize, without however offering himself as the site of a living meditation aimed at realising that desire, which would mean, above all: suffering and thinking.

Of course, one doesn't need to think of the devil to imagine that truth can translate into deception; there are periods and circumstances in which all expressible truths are deceptive. That is why we are here arguing about the unthought.

When we say that 'power and politics are not the same thing', I am certain that we say something true. By *saying the truth*, I mean that we interpret reality in the sense of saving it from 'coming to nothing'—and I add, without taking this further, that this 'coming to nothing' is not a figure of speech, but tends to embody itself in actual behaviours. Nihilism is not only philosophy, and it is not possible to say the truth if this truth is not, somewhere, in some way, at work or if it has really not begun to act.

That power's embrace is fatal for politics is well known, it's not my discovery. Power transforms those who believe they possess it into its cogs. It would not be difficult to demonstrate that, among men of action as well as political thinkers, the most brilliant ones are those who oriented themselves, more or less consciously, precisely in the direction of holding political life back from being devoured by the logic of power. We call these men of politics, the others are men of power: I am thinking of the difference between Aldo Moro and Giulio Andreotti, but I am also thinking of the great political thinkers, like Machiavelli himself, thinkers who should be seen in this light as inventors of politics, there where before there was only the flat logic of power.

The logic of power can be summed up in two points: one says that power always holds the sword by its hilt, that it cannot stand the distinctively human experience of vulnerability; the other says that power uses everything and everyone, even those who have it. Many years ago, Giulio Andreotti made an ironic quip about the saying according to which 'power ruins [*logora*] those who have it', countering that it ruins those who *do not* have it. But he himself has become the living proof of what power can do to those who have it, as is brilliantly shown by the director Paolo Sorrentino and the actor Toni Servillo, respectively the author and principal interpreter of the film *Il Divo* (Italy 2008)

So, though I am sure that we are speaking the truth, I am still stuck with Banquo's suspicion: that the devil too can speak the truth and that I may deceive myself. There is a fake politics, more common among women than men, which is based on taking a safety distance from power. I object to it that politics indeed cannot agree with power, but that the latter's pressure cannot be ignored. Power is something that 'presses', in the sense of the Italian word *premere*: it both oppresses and attracts. To make power into the *raison d'être* of politics is aberrant; to keep it at a safety distance is illusory.

What does this mean in practice? Does it all come down to finding the right middle path between two extremes? Or can we, must we, find a less obvious and more incisive articulation?

From the exchanges that followed my first contribution on this theme (in the journal *Via Dogana*, 86, September 2008), there emerged the suggestion that we should enter into the devil's territory, to wrest reality from it. What I mean with this colourful expression, which I stole from the writer Flannery O'Connor, is that we need to know how to give up any truth, even the dearest or most solid, to render speakable what the dominant discourse, even in our head, has silenced and which, because of this muteness, makes our experience insipid and our reality unreal.

The quarrel cannot fail to also, and perhaps above all, concern the word 'power'. This is an extremely important word, whose use however has been concentrated in a meaning closely related to domination. In her political study regarding violence, *On Violence* (1970), Hannah Arendt quotes Voltaire's definition 'Power consists in making others act as I choose', and those of others, among whom Max Weber: there is power anytime I can 'assert my own will against the resistance' of others.[3] Given these definitions, with which she's not satisfied, Arendt lets a positive meaning speak, a meaning which is untied from domination and subjugation. The risk here becomes that of ambiguity, and perhaps we need to run it, in order to chase our adversary to the very end.

Neither definitions nor terminology will help us escape the confusion that takes place within and without us, without any clear boundary, between power and politics. We need to distil,

---

3. Hannah Arendt, *On Violence*, New York, Harcourt & Brace, 1970, p. 36.

as I already said, that is we need to put an end to confusion by changing the relation between ourselves and the world, which in practice means putting ourselves in a situation, like the self-awareness group in the women's movement, in which it is possible to experience freedom (that is, to go to the roots of freedom) and to base the very idea of politics in this changed relation and this new experience.

Accordingly, the answer cannot be sought without running the risk of deception and ambiguity. I know women and men who have this courage, who make no concessions to ideology or dogmatism, but too often it happens that if they trip up, something that easily happens in the devil's territory, they don't want to recognize that they made a mistake. If this is the case, it's a thousand times better to stay out of it!

To think the unthought means doing without the criteria that would be indispensable for a secure judgment. However, by exposing ourselves to the risk of self-deception, of not being understood or losing our way, it happens that we make formidable discoveries, as long as we don't lose our awareness of this exposure. Flannery O'Connor would have said: as long as we don't forget that there is the devil and then there is grace.

I want to stress that to speak as I have, of the devil's territory, does not mean demonising power, and to speak of entering this territory is not an invitation to make compromises with it. The outcome can be something entirely different, and that is my aim.

Take the case of Freud: making his first uncertain steps in what would then become the theory and practice of psychoanalysis, he listens, from his first women patients, to repeated tales of sexual abuses suffered in childhood at the hands of members of the family, fathers not excluded. He is very struck by this, and to some extent troubled, but, since he's looking for a cause of his patients' illness, he does not recoil and perseveres. The doubt, however, stays with him—for reasons which are in part mistaken, it should be said (can fathers really do such things?)—pushing him into an unprecedented direction, where there is neither truth nor falsity, neither good nor evil. This will prove to be the right direction. In brief, Freud comes to understand that the patient, with her tale of seduction, has put into words and communicated to him a world which otherwise could not be put into words. Today we tend to think that these tales of sexual abuses in the family were veridical in a realist sense too, but this consideration does not invalidate Freud's move, to abandon the judgment about truth and falsity at the level of reality in order to shift to another level and become the ear listening to the unconscious. In other words, to become the living mediation in the place of the absent mediations.

A contradiction rears its head here: it seems that, by exposing ourselves to the risk of deception and ambiguity in order to rescue reality, we are led to some kind of independence from reality as a solution. That is precisely how it is: but we are dealing with a symbolic independence. Not an avoidance of reality, but a way of being in it without absolutising it. It is not even a matter of bracketing reality, which would be an idealist solution, since such a bracketing is not within our power. It is in our power to 'suffer and think', that is to remain in the state of impotent desire, to pierce the horizon in which the real is inscribed and takes on this or that name. This allows us to think without names and to find ourselves in the place where the unthought happens.

This idea complicates our framework: the struggle against unreality requires something like a suspension of reality's dictates. But it also helps us to broach the theme of the disjunction between politics and power without demonising power, on the one hand, and without reducing ourselves to the rhetoric of dirty hands—on the other, a rhetoric which is generally used to justify the confusion between power and politics.

The dictates of reality are to some extent always complicitous with power, as Foucault maintained and illustrated. The 'ignorant' people always knew this, and in the end positivist philosophers discovered it too. Some of them, in the Vienna Circle, strained to define the dictates of reality in their pure state, without succeeding. Ernst Mach, their initiator, had no doubts that knowledge is always, to some extent, an interpretation: dictated by whom? In other words, we are not strictly speaking dealing with a symbolic independence from reality (which would be madness), but from power.

The symbolic independence from power, which I have already said is the agent of the undoing of the embrace between politics and power, is more than a moral virtue, and more than extraneousness from power. I don't know anything more radical in our tradition, when it comes to symbolic independence, than the following passage from Saint Paul's Epistle to the Romans, whose radicality is such as to almost make what is at stake unrecognisable. This is what the apostle who wasn't one says:

> Do not be overcome by evil, but overcome evil with good. Let every person be subject to the governing authorities; for there is no authority except from God, and those authorities that exist have been instituted by God. Therefore whoever resists authority resists what God has appointed, and those who resist will incur judgment. For rulers are not a terror to good conduct, but to bad. Do you wish to have no fear of the authority? Then do what is good, and you will receive its approval; for it is God's servant for your good. But if you do what is wrong, you should be afraid, for the authority does not bear the sword in vain! It is the servant of God to execute wrath on the wrongdoer. Therefore one must be subject, not only because of wrath but also because of conscience. For the same reason you also pay taxes, for the authorities are God's servants, busy with this very thing. Pay to all what is due them—taxes to whom taxes are due, revenue to whom revenue is due, respect to whom respect is due, honour to whom honour is due. Owe no one anything, except to love one another; for the one who loves another has fulfilled the law. The commandments, 'You shall not commit adultery; You shall not murder; You shall not steal; You shall not covet'; and any other commandment, are summed up in this word, 'Love your neighbor as yourself'. Love does no wrong to a neighbor; therefore, love is the fulfilling of the law (Rom. 12, 21-13, 8).

I won't dwell on the philological issues raised by this famous passage on obedience to the constituted authorities, which are not such as to hinder its insertion into the context of my reflection; I would simply like to note that, in my quotation, the Pauline passage is framed by two phrases that recall its context, whose theme is love for others (*agape*). In this, I follow Karl Barth's authoritative teaching.

At bottom, the problem posed by this passage from the Epistle to the Romans does not lie in its literal meaning, which is clear. The problem is another one, which is that it seems unacceptable to very many people because it appears to deny any value to the endeavour to change the order, or more often the disorder, of this world.

Whoever reads Paul with faith and fervour is tempted at this point to introduce some reasonable interpretation, to make him acceptable to our culture, or to posit a definitive historical distance. I think we should not make efforts in this direction, which would simply weaken the meaning of Paul's lines. They should be left in their literal meaning: the more astonishment they elicit, the better. The radicality of the solution that Paul teaches to the community of Christians of Rome is in fact such as to flip into its opposite. A veritable riddle. The astonishment increases if we recall that the writer is not a man of order; on the contrary, he was someone who was consciously defying the dominant civilisation, and those words are an integral part of the challenge. In light of this reflection, the text ultimately appears as a *cryptogram*, whose meaning is at first sight entirely opaque, but which becomes obvious as soon as you grasp it.

The meaning which I glimpsed, as I have already remarked, is the teaching of the symbolic independence from power and the subtraction of oneself from its grip by eliminating any obligation or expectation in its regard. At the heart of the Pauline teaching lies the invitation to not chain ourselves to the level of the relations of forces by opening lines of credit with respect to the constituted authorities. Therefore, it's better to give to them everything they demand and not believe that opposing them would produce something that is other from the mere repetition of the evil which we are trying to combat. Remember the beginning and end of the passage, which situate themselves on another level of being, where those baptised in Jesus Christ now live, the level in which evil is fought with means that it is radically ignorant of, the level of love which is the only debt towards others. Beyond reactive opposition and revolutionary rebelliousness.

In his famous commentary to the Epistle to the Romans, Karl Barth says: the men of power who serve the order of this world will be punished by the revolt of the poor and they will receive their judgment historically; not so the revolutionaries, who are better and who fall into an error that no one rescues them from, because their defeat is the punishment of the dominant power, the most deceptive. That is why it is their error that must be corrected and that is why it is to them that Paul speaks.

We can find something both very different and very similar in the text of Hannah Arendt we've already mentioned, where she speaks of revolutionary violence, which mistakenly claims to be able to interrupt the repetitive course of human history.

> If we look on history in terms of a continuous chronological process ... violence in the shape of war and revolution may appear to constitute the only possible interruption. If this were true, if only the practice of violence would make it possible to interrupt automatic processes in the realm of human affairs, the preachers of violence would have won an important point. ... It is the function, however, of all action, as distinguished from mere behaviour, to interrupt what otherwise would have proceeded automatically and therefore predictably.[4]

As is well known, Arendt disagrees with those who theorize that politics is nothing but the struggle for power, and is even less sympathetic to the argument that violence is the quintessence of power. She proposes the following reflection, which has inspired me and many others in our research: 'It is, I think, a rather sad reflection on the present state of political science that our terminology does not distinguish among such key words as "power", "strength", "force", "might", "authority", and, finally, "violence"— all of which refer to distinct, different phenomena and would hardly exist unless they did. ... To use them as synonyms not only indicates a certain deafness to linguistic meanings, which would be serious enough, but has resulted in a kind of blindness with respect to the realities they correspond to'.[5] But she herself, as I've already noted, does not dispel the ambiguity of the word 'power' and speaks of a 'living power', originating in the concerted political action of many (her model, as is well known, is the democracy of 'councils', which she takes from Rosa Luxemburg). The symbolic break which the Pauline text articulates as fighting evil with good, is analogously given in Arendt by the explosiveness of an acting in concert that makes something happen in the order of freedom, and which replaces revolutionary violence by undermining the power of power, leaving it naked.

The thesis that we're advancing is sufficiently clear: politics untangles itself from power thanks to its symbolic independence with regard to power itself. Or better, politics is the name for our slipping away or untying ourselves from the tangle of the relations of force by which we are molded and restrained, in order to shift to

---

4. Arendt, *On Violence*, pp. 30-1.
5. Arendt, *On Violence*, p. 43.

another level, in a move whose character is distinctively symbolic.

The terms of the disentanglement (the *dénouement*, to borrow a fine French word) from the embrace between religion and power, which was immensely strong in the Roman Empire, such as they were thought and proposed by Paul to the community of the Christians of Rome, were reprised in our epoch by thinkers like Karl Barth and Simone Weil, authors who have continued to lend illumination and aid to much contemporary thought.

We are therefore dealing with a vision which has not been entirely forgotten, even if we leave aside its secular rediscovery by Arendt. But we know that Christianity, after having defeated the political theology of the Roman Empire, to a certain extent restored it. The Pauline idea of the symbolic independence from power functioned historically, but it later became the proverbial ladder which is thrown away once one gets to the top.

What's the point then for us (me) to bring it out again? My answer is brusque: it is not us, it is not me, it is the idea that forces itself upon us. The idea has reared its head with the historically unpredictable fact of women's freedom and it cannot be chased back, because it has the strength to open up a horizon which had been closed down.

Translated by Alberto Toscano

8 TOWARDS A CRITIQUE OF POLITICAL DEMOCRACY

Mario Tronti

A word of warning: my argument will involve a deconstruction of the theme of democracy. I will seek to clear the field of the conceptual debris that has accumulated around the idea and practice of democracy, so that our discussion can then take up—in a more constructive and also more programmatic manner—the identification of further directions of inquiry, especially in what concerns that crucial passage represented by the construction of the subject.

I believe that the moment has really come to undertake a critique of democracy. These moments always come. They come when the objective conditions of the matter at hand meet with the subjective dispositions of the one who confronts and analyses it. A trajectory of thought has developed on this terrain, which I believe is today capable of grasping the crisis of an entire practical and conceptual apparatus. That is because when we say democracy we say this: *institution plus theory; constitution and doctrine*. A very powerful bond was established among these terms, what we could even call a knot. This knot does not just bind together socio-political structure and strong traditions of thought (those of democracy are always robust intellectual traditions, even if the current drift in the practice of democracy suggests the presence of a weak terrain); it is internal to the practical structures and the traditions of thought themselves. That is because within democracy, within its history, we find knotted together a practice of domination and a project of liberation—they always present themselves together, they are co-present. In some periods (periods of crisis, states of exception) these two dimensions are in conflict. In others (such as in the contemporary situation, which is a state of normality, or at least that is the way I read it) they are integrated. And these two dimensions—practice of domination and project of liberation—are not two faces; they are the single face, a *Janus bifrons*, of democracy. Depending on the way that the balance of forces between the top and the bottom of society is established, articulated, and constituted, sometimes one is more visible than the other. I think that at this juncture the balance of forces is so weighed to one side—the side hostile to us—that we can only see a single face. This is the reason why democracy is no longer the best of the worst; it is the only thing there is. That is, there is nothing else outside it.

Now, if this is the knot, while in the past we attempted (or at least I attempted) to untie it, I think the moment has come to cut it. This requires a new configuration of the critique of democracy, which thereby assumes a very radical character. The determinate critique of democracy that I am advancing here has a father, *workerism*, and a mother, *the autonomy of the political*. And it is a female offspring because the thinking and practice of *difference* have anticipated this critique with the questioning of the universalism of the *demos*—which is the other face of the neutral character of the individual—and with that 'don't think you've got any rights' which is no longer addressed to the single individual but to the people. There is in democracy an identitarian vocation hostile to the articulation of any difference whatever as well as to any order of difference. Both the *demos* and the *kratos* are unique and univocal, rather than dual, entities; they are not and cannot be split. Democracy, as is widely known, presupposes an identity between sovereign and people: sovereign people, popular sovereignty, so goes the doctrine. During a long phase of modernity, in the nineteenth and especially the twentieth century, this identity of sovereign and people has been answered by a kind of spirit of division stemming from a society split into classes. Obviously, this was a raw indication of the ideological falsity at the heart of such an identity. Or rather, it put the very conceptual structure underlying the identity into crisis. So it was that during this phase the very separation of powers—within an apparatus that attempted the great passage from liberalism to democracy, and then the conjugation of liberalism and democracy—revealed itself precisely as a mask, the mask of the unity of power in the hands of one class. I believe that it is from here that we must start again in order to follow, genealogically, the trajectory of the accomplishment of democracy, in the passage from thought to history. My perception is that from its origin, this practical concept, this theoretical-practical knot that is democracy, unravels towards the conclusion that we are living through in this phase; so much so that the democracy of the moderns, considered both in its principles and its realizations, can now be judged by its results.

I speak of real democracy in the same sense that it has been possible to speak of real socialism. Real socialism did not indicate a particular realisation of socialism that left open the possibility of another socialism, the ideal one. For socialism incarnated itself in that realization to such an extent that at this point, in my view, 'socialism' is what took place there and then, and nothing else. There is no possible recuperation of the symbolic order that was evoked by this word; it is not possible to detach it from the reality that embodied it. The same I think can be said of contemporary democratic systems, which should not be read as a 'false' democracy in the face of which there is or should be a 'true' democracy, but as the coming-true of the ideal, or conceptual, form of democracy. In this case too, it is impossible to save this concept from its effective realization. And as I remarked above, contrary to what is commonly thought today, it is not in its past or in its theories but rather in this realisation that democracy has become a weak idea, to the point that 'democracy' is a noun in constant need of qualifying adjectives. When a noun needs adjectives in order to define itself, it is a sign of a lack of conceptual autonomy. Today in fact we say liberal democracy, socialist democracy, progressive democracy; some have even spoken of totalitarian democracy, and so on: all elements that point to a weakening of the concept.

At this point I must warn you that in this critique of democracy I am not retracing the gestures of what has been defined as the critique of totalitarian democracy. If anything, I am using the liberal critique of democracy—Locke versus Rousseau and so on—together with the important twentieth-century elaborations that follow in this tradition: Hayek's work is a salient example. The long, or rather not long but intense age of totalitarian or authoritarian solutions really made possible the definitive victory of democracy. Germany and Russia, in my view, bear the historical guilt of letting America win precisely through those solutions that served to reinvigorate the solution provided by American democracy.

Democracy has problems with freedom. If it is true that real democracy is configured as liberal-democracy and that in the end this has been the winning solution, it is precisely this conjunction, binding together freedom (or liberty) and democracy, that must be critically attacked. It is a matter of detaching and juxtaposing the two terms—freedom *versus* democracy—because democracy is identity to the same extent that freedom is difference. The problem of democracy must then be confronted on two sides: a deconstructive critique of democracy must be accompanied by a constructive theory, what I would call a foundational or re-foundational theory of freedom, of the concept and practice of freedom. As we elaborate the figure of the subject, we should keep in mind that the subject needs to retrace the form of freedom. Because it is precisely difference that is the foundational element of freedom and the dislocating element of democracy.

As you will be aware, I move within a framework that I ironically refer to as *neo-classical,* in the sense that I place myself in the twentieth century. I plant my feet in that century and from there I look backwards and forwards. I have no intention of moving from there. So it is that the authors that I keep coming back to with regard to this theme are Kelsen and Schmitt, who strangely, in the same period (Kelsen in 1929 with *Democracy* and Schmitt in 1928 with *Constitutional Theory*), despite being divided in everything else, are fundamentally united in the critique of democracy, or rather in the unveiling of the democratic enigma. Kelsen says:

> The discord between the will of the individual—the starting point of the demand of freedom—and the order of the state, which presents itself to the individual as an external will, is inevitable. The protest against the domination of someone who resembles us, leads in political consciousness to a displacement of the subject of domination which is also inevitable in the democratic regime, that is, it leads to the formation of the anonymous person of the state. The

imperium derives from this anonymous person; not from the individual as such, but from the anonymous person of the state. The wills of the single personalities give free rein to a mysterious collective will and a collective person which could even be characterized as mystical.

Schmitt makes analogous considerations, when he says:

> Democracy is a state-form that corresponds to the principle of identity; it is the identity of the dominated and the dominating, of the governing and the governed, of those who command and those who obey. And the word 'identity' is useful in the definition of democracy because it points to the complete identity of the homogeneous people, this people that exists within itself qua political unit without any further need for representation, precisely because it is self-representing.

It is with regard to this self-representation that democracy becomes an ideal concept, because it indicates, as Schmitt says, 'everything that is ideal, everything that is beautiful, everything that inspires sympathy. Identified with liberalism, with socialism, with justice, humanity, peace, the reconciliation among peoples and within the people'. 'Democracy'—as Schmitt remarked in another fine sentence—'is one of those dangerous complexes of ideas in which we can no longer make out concepts'. This then is the democratic enigma.

The focus is therefore democracy not as a form of *government* but as a form of *state*: that thing that took the name of *democratic state*, which evolved on the basis of the nineteenth-century coupling of the workers' revolution and the great crisis, a decisive coupling for the subsequent history of capital and for the manner in which capital exists today at the global level. Through the social or welfare state we have witnessed a gradual process of extinction of the state, which obviously is not complete but which is quite advanced in this phase, and which has been accelerated by all the processes of globalisation. Moreover, the analysis of the network of global domination confirms this passage: the extinction of the state in democratic society; the recuperation of the function of the state within the social. It is here that we encounter an essential shift, because politics, in my view, comes to be accomplished not institutionally but sociologically. And it is democratic society that has resolved the contradiction in the terms harboured by the concept and practice of the democratic state. Thus we have seen the passage of democracy from a form of government, in the democracy of the ancients, to a form of the state, in the democracy of the moderns, to a form of society, in the twentieth century.

I feel I can advocate the thesis that capitalism, as it develops, becomes ever more and ever more successfully *bourgeois society*. It is not correct to say that we have overcome the bourgeois character of society; we could even say that it has finally been achieved. Bourgeois society seems a dated, passé term, but in my view it is once again extremely timely. Precisely in the sense that society started as *bürgerliche Gesellschaft*, that is, simultaneously as civil society and bourgeois society. The entire recent history of the twentieth century—after the 1970s of the movements and of feminism, and all the vicissitudes of the response to that moment—can be read as a recuperation of capitalist hegemony through the return of the figure of the bourgeois. So much so that the distinction-juxtaposition of *bourgeois* and *citoyen* is rescinded, as the latter comes to be recuperated by the former. We witness the epochal encounter between *homo oeconomicus* and *homo democraticus*. The subject of the spirits of capitalism is precisely the *animal democraticum*. The figure which has become dominant is the *mass bourgeois*, which is the real subject internal to the social relation. There will be no genuine and effective critique of democracy without a profound anthropological investigation, a social anthropology but also an individual anthropology, taking 'individual' here too in the sense of the thought-practice of difference.

Here we must give great importance to both the imaginary and the symbolic. Much hangs on this, as can be seen in the return of the myth—coming to us from the United States—of *the society of owners*. It comes precisely from the America of Bush and the neo-cons, from this interesting episode of conservative revolution that is taking place there and that we should keep under watch. After all, democracy is always 'democracy in America'; and the United States has always exported democracy with war. We are stunned that they are doing so now, but they have always done so. They even brought it to Europe through the great wars. The allied armies did not liberate us: they democratized us. In fact, it is after the age of the European and world civil wars that democracy truly triumphed. And democracy was finally decisive for the victory of the West in the last war, the Cold War.

Contrary to what one often hears, especially from progressive quarters, I deny that in the current phase we are experiencing the centrality of war. It seems to me that this present emphasis on peace-war is entirely disproportionate. All the wars are taking place at the borders of empire—on its critical fault-lines, we could say—but the empire is internally living through its new peace, though I do not know if it too will last one hundred years. It is in this condition of internal peace and external war that democracy does not merely prevail, but experiences a resounding triumph. In order to understand its power we must define its mass base. Democracy today is not the power of the majority. It is, as we were trying to suggest through the categories of identity and of the homogeneous people, the power of all. It is the *kratos* of the *demos*, in the sense that it is the power of all on each and every one. That is because democracy is precisely the process of the homogenization, of the massification of thoughts, feelings, tastes, behaviours expressed in that political power which is common sense. Common sense, when it becomes the property of a mass and meets with good sense, constructing this symbolic democratic order, verifies to some extent what Marx said when he argued that theory becomes a material force when it takes hold of the masses: common sense also becomes a material force when it takes on a mass dimension. It is important to note that this mass establishes and unifies itself not around goods as much as around values, and it is this form of mass that we must be able to define, so as to then understand how it can be undone. At least the body of the king was double, as the great interpreters taught us, because there was still a sacralization of power. Now instead, with the secularization of power, the body of the people is single, univocal. The processes of secularization have had a huge influence on these types of issues. A critique of secularization still stands before us as something we have yet to confront and carry out.

Basically, I see a kind of *mass biopolitics*, in which singularity is permitted for the private but denied to the public. The 'common' which is spoken of today is really that *in-common* which is already wholly taken over by this kind of self-dictatorship, this kind of tyranny over oneself which is the contemporary form of that brilliant modern idea: voluntary servitude. After the twilight of the glorious days of class struggle, we have not seen the victory either of the great bourgeois—the one *à la* Rathenau who we liked so much when we were young—nor the petty bourgeois who we always hated. The average bourgeois has won: this is the figure of democracy. Democracy is this: not the tyranny of the majority, but the tyranny of the average man. And this average man constitutes a mass within the Nietzschean category of the *last man*.

Of course, I am radicalizing these shifts, in part because that is how I am used to thinking—i.e. radicalizing problems—and also because I am trying to understand the astounding *silence of revolution* in these decades, in this phase. This is what I am trying to shed light upon, this darkness. Years ago, you could read the following Marxian lines under the masthead of *Classe Operaia*: 'the revolution is still going through purgatory...'.[1] Well, what effectively happened is that there was no passage to paradise, but rather, I would say, a descent into hell.

Democracy is antirevolutionary because it is antipolitical. There is a process of depoliticization and neutralization that pervades it, impels it, stabilizes it. And in my view this antipolitics of democracy is the point that I take as the offspring of that entire phase which I referred to as the autonomy of the political. What is more, I read this datum empirically in the conquest and management of consensus with which, when all is said and done, contemporary political systems are in practice identified. I don't call them political systems any longer, but *apolitical systems*. Western society is no longer divided into classes, in that antinomy of the past, but into two great aggregates of consensus, of equal quantitative consistency: in all Western countries this consensus, from the United States to Italy, when the votes are tallied up, ends up being 49 to 51, or 48 to 52. Consensus, thus, is divided in two. Why? Because on the one side we have reactionary

---

[1]. *Classe Operaia*, a 'political monthly of workers in struggle', was published, under Tronti's editorship, between January 1964 and March 1967, when it broke up due to political differences in the editorial board. Its first editorial, 'Lenin in England' (later collected in Tronti's *Operai e capitale*), formulated the fundamental workerist thesis, according to which the working class and its struggles came first, and capital and its development should only be considered as a consequence of and reaction to these struggles. Among the contributors to *Classe Operaia* were Antonio Negri, Romano Alquati, Sergio Bologna and Ferruccio Gambino. [Translator's note.]

bourgeois drives, and on the other progressive bourgeois drives. And I say drives, that is, emotive reflexes, symbolic imaginaries, all moved and governed by great mass communication. Reactionary and progressive drives which nonetheless share this average bourgeois character. On the one hand compassionate conservatism, on the other political correctness. These are the two great blocs. This is the governmental alternative offered by apolitical democratic systems.

In this condition there is no possibility either to be or to make a majority. We must remain in the condition of a strong and intelligent minority. For some time, without great success, I have argued for the necessity of revisiting the great theoretical moment of the *elitists*.[2] I get no further because the resistances—which here too are both emotional and intellectual in character—are strong. But the elitists were the only ones to have formulated a critique of democracy before the totalitarianisms. And if that critique of democracy had been kept in mind, perhaps a correction of the democratic systems would not have allowed the age of totalitarianisms. The elitists' critique of democracy was not made from the point of view of absolutism. On this point the lineage instead comes from workerism. Let me clarify this otherwise opaque affirmation. Mulling it over, I have come to the conviction that the working class was the last great historical form of social aristocracy. It was a minority in the midst of the people; its struggles changed capitalism but did not change the world, and the reason for this is precisely what still needs to be understood. But what can already be grasped is how the workers' party became the party of the whole people, and how workers' power, where it existed, became the popular management of socialism, thereby losing its destructive antagonistic character. And this was one, if not the only, element that made possible the workers' defeat.

Let me conclude. I do not know if the multitude can be understood as a mass aristocracy—if that were the case, then these arguments would in some sense converge and this deconstructive operation could allow us to leap to a higher level. But I also know that if the conditions that I have described remain, the sub-

---

2. Tronti is alluding to the sociological works of Vilfredo Pareto (*The Rise and Fall of Elites*), Gaetano Mosca (*The Ruling Class*) and Robert Michels (*Political Parties*), among others. [Translator's note.]

ject is entangled in this web. If the multitude remains caught up in the web of really-existing democracy I do not think it will be able to definitively escape the very web of neo-imperial power. A contemporary feature of Empire is in fact that it is a *democratic Empire*. If these conditions are not put into crisis, the subject itself cannot manage any effective political manoeuvre in this situation, through an alternative network, for the sake of another possible historical break.

Translated by Alberto Toscano

9 CHRONICLES OF INSURRECTION: TRONTI, NEGRI AND THE SUBJECT OF ANTAGONISM

Alberto Toscano

> Once I went to May Day. I never got workers' festivities. The day of work, are you kidding? The day of workers celebrating themselves. I never got it into my head what workers' day or the day of work meant. I never got it into my head why work should be celebrated. But when I wasn't working I didn't know what the fuck to do. Because I was a worker, that is someone who spent most of their day in the factory. And in the time left over I could only rest for the next day. But that May Day on a whim I went to listen to some guy's speech because I didn't know him.
> Nanni Balestrini, *Vogliamo tutto*

> Force ... is itself an economic power.
> Karl Marx, *Capital, vol. 1*

BEFORE EMPIRE, BEHIND THE MULTITUDE

Though much work has been carried out to rectify, whether critically or affirmatively, a dehistoricized understanding of the political content of the theses forwarded in books like *Empire* and *Multitude*, there remains a strong tendency—at times enabled by their own rhetoric of rupture and transformation—to treat the recent works of Hardt and Negri as a kind of theoretical UFO, or better a time-machine emancipated of all nation and class coordinates, visiting us from a vibrant future that the authors insist in describing as our present. Behind the seemingly apologetic and impressionistic character of the figures of Empire and the multitude lies a long, punctuated history of theoretical work and political practice aimed at testing the validity of Marxist categories in light of empirical transformations in modes of production and reproduction, tendencies in class composition and shifts in the forms of capitalist domination, driven by political struggles and economic reconfigurations in post-war Italy.[1] Behind the non-dialectical pairing of Empire and multitude, one needs to discern the figures of a far more classical albeit 'mutant' antagonism between capital and labour, of the kind formulated in what can loosely be defined as the 'workerist' (*operaista*) and 'post-workerist' (*post-operaista*) development of critical Marxism beginning with the work of Raniero Panzieri and the *Quaderni Rossi* journal, and then gaining greater prominence chiefly in the writings of Mario Tronti and Antonio Negri, whose intellectual production of the sixties and seventies will concern me here.

My aim in this article is to explore the following question: What drives the move from the 'workerist' dialectic of antagonism and its capture, through the insurrectionary unilaterality of worker's autonomy, all the way to the recent theories of exodus? In order to sketch an answer to this question, we need to investigate the juncture between the political-economic logic of capital and the revolutionary logic of separation—of *communism as separation*.[2] In the epoch of what Marx referred to as 'real subsumption', wherein all labour and production processes take place within the ambit of capitalist relations, it is only an organized act of antagonistic separation that, from the vantage point of *operaismo*, can elicit the emergence of living labour as a collective subject capable of appropriating a production process founded on the exploitation of its capacities. As Negri remarks, capitalist 'totality is a texture in which we find ourselves and in which we must separate ourselves in order to exist—but it is the intensity of the separation, the force with which antagonism is recognized, that

---

1. See Guido Borio, Francesca Pozzi and Gigi Roggero, *Gli operaisti*, Roma, DeriveApprodi, 2005, Steve Wright, *Storming Heaven: Class Composition and Struggle in Italian Autonomist Marxism*, London, Pluto, 2002.

2. For a treatment of this concept—which also features strongly in Negri's 'The Italian Difference'—with reference to the work of the French philosopher Alain Badiou, see Alberto Toscano, 'Communism as Separation', in Peter Hallward (ed.), *Think Again: Alain Badiou and the Future of Philosophy*, London, Continuum, 2004.

constitutes us as singularities—as subjects'.[3] The open paradox of the workerist 'tradition' (to adopt a term whose intensely problematic character has been highlighted by Sergio Bologna) and of the political philosophy of the multitudes that has followed in its wake—which is of course a paradox faithful to some of the key insights of Marx—is precisely the twin affirmation of an integral *immanence* of capitalist relations to the social (of a thoroughgoing *socialization* of production) and of the radicalization of the antagonism between capital and labour. Subsumption, precisely to the extent that it is real, manifests itself as an irrational form of command and heralds the possibility of a communist appropriation of production. In a nutshell, the problem is that of the realization of communism in a situation of advanced and dynamic capitalism, in which political crisis and antagonism are by no means necessarily accompanied by scarcity or stagnation (as witnessed by the fact that the golden age of FIAT in Italy was concurrent with fierce struggles that invested the factories themselves, whilst the relative social peace of the 80s and 90s saw its progressive enfeeblement and eventual collapse).

TENDENCY AND COMMUNISM

Such a position rests on the conviction that the rule of capital is now divested of any possibility for mediation, dialectics, or measure. It posits the rupture, catalysed by worker's struggles, of any social-democratic, Rooseveltian, or Keynesian project. However, and this point is paramount if workerism is to include its own 'refutation of idealism', the putative collapse of measure and mediation must itself be the outcome of a historical process. It must itself be the product of a dialectic—albeit a dialectic that seems to signal the impossibility of further dialectical mediation. In Negri, it is the concept of *tendency* that provides this historical determinacy, rather than that of a closed and endogenously developing dialectical totality. Negri defines it as follows:

> The tendency gives us a forecast that is determinate, specified by a materialist dialectic which is developed by the factors comprising it. The tendency is the practical/theoretical process whereby the working-class point of view becomes explicit in its application to a determinate historical

3. Antonio Negri, *Fabbriche del soggetto*, Livorno, XXI Secolo, 1987, p. 224.

epoch. This means that to pose the tendency, to describe it and to define its contradictions is a far cry from economic determinism. Quite the opposite: to pose the tendency is to work up from the simple to the complex, from the concrete to the abstract, in order to achieve an adequate overall theoretical perspective within which the specificity and concreteness of the elements which were our initial starting point may then acquire meaning. [...] [It is] reason's adventure as it comes to encounter the complexities of reality.[4]

Without such a concrete tendency, communism would be reduced to the unilateral purity and impotence of a terroristic decisionism incapable of intervening in the real articulations of systemic development. Viewed in this light, workerism, as the militant combination of political-economic forecast and organized intervention, can serve as a useful corrective to the dominant perception of Marxism as first and foremost a theory of systemic transformation, one that necessitates supplementing by specifically 'political' theories of antagonism, hegemony and subjectivation. The workerist gambit—later radicalized in the theories of workers' autonomy and self-valorization—lay in arguing that one can move beyond a treatment of the dynamic of capitalism solely in terms of exploitation and the vampire-like 'absorption' of living labour as variable capital into the process of production, to a consideration of the driving importance of the subjectivity and organization of the working class, shifting analyses of the transformations of capitalism firmly onto the level of a materially and temporally determinate antagonism. In other words, workerism revitalizes the Marxian thesis whereby the parameters of the capitalist domination and exploitation of labour-power and the extraction of surplus-value are political through and through. As we shall see, this is not simply a theoretical posit, but is accompanied by an analysis of the politico-economic conjuncture via the prism of the tendency. From this perspective, according to Negri, the complex mediations of the law of value that had played such a dominant role in the American New Deal, in the fortunes of Keynesianism and in the entire tradition of social democracy become increasingly obsolescent, as capital manifests itself increas-

---

4. Antonio Negri, *Revolution Retrieved: Selected Writings on Marx, Keynes, Capitalist Crisis and New Social Subjects 1967–1983*, trans. Ed Emery and John Merrington, London, Red Notes, 1988, p. 125.

ingly as a form of political command desiring ever greater autonomy and ever-diminishing negotiation with labour power.

The thesis of Negri and his comrades at the time was that such an 'autonomization' of capital—marked by an increasing reliance on monetary, fiscal, and financial policies to the detriment of social planning, as well as by the concomitant forms of enforcement and control—can be regarded as the *effect* of an ever-greater claim to autonomy and self-determination exerted by working-class struggles to appropriate a domain of production and reproduction which, far from being relegated to the factory alone, now covers the entirety of the social fabric. Though the concepts of 'class composition' first and of 'organized autonomy' later mark the sensitivity of this approach to the complexity and power dynamics of antagonism, we could still say of 'antagonism in general' what Marx says of 'production in general'; to wit, that it is 'an abstraction, but a rational abstraction insofar as it really brings out and fixes the common element and thus saves us repetition'.[5]

The question of workerism—and then of autonomism and post-workerism broadly construed—was that of how to perpetuate, at the level of political strategy and organization, the idea of communism as the suppression of work. In other words, how to effect a practical transition to communism in the conditions of a highly socialized economy but also one characterized by a high dose of political repression. It is in this sense that we should grasp the three theses that Negri posits as crucial to his politics of antagonism: 'all Marxist categories are categories of communism';[6] 'communism has the form of subjectivity, communism is a constituting praxis';[7] 'communism is in no way a product of capitalist development, it is its radical inversion'.[8] Evidently, the principal theoretical enemy here is any variety of (parliamentary) socialism, to wit any attempt to think the suspension of capitalist relations as a possible result of a mediation organic to the capitalist mode of production—be it as a 'natural' out-

---

5. Karl Marx, *Grundrisse: Foundations of the Critique of Political Economy*, trans. Martin Nicolaus, London, Penguin, 1993, p. 85.

6. Antonio Negri, *Marx Beyond Marx: Lessons on the Grundrisse*, Jim Fleming (ed.), trans. Michael Ryan, Mauricio Viano and Harry Cleaver, New York, Autonomedia, 1989, p. 161.

7. Negri, *Marx Beyond Marx*, p. 163.

8. Negri, *Marx Beyond Marx*, p. 165.

come, as the progressive accumulation of victorious reforms, or as the gradual effect of the shows of force of the working class and its party leadership. Against any such faith in mediation, Negri wishes to affirm the 'antagonistic nature of Marxist logic'. As he writes: 'The antagonism must become social, *global revolutionary power must become a revolutionary class* against capitalist development'.[9] Such affirmations cannot fail to trail a whole set of thorny questions in their wake. To begin with: What is the nature of the purported independence of the proletariat? Does it possess a kind of social latency or is it a product of political will and organization, *ex nihilo*? How can we think the political and programmatic autonomy of the exploited, as well as the full immanence of the antagonistic class within capital? In other words: What is an immanent antagonism, a separation in and against real subsumption? It is only by confronting this question that I think light can be shed on the practical-historical shortcomings and theoretical potential of workerism and autonomism, as well as upon the antagonistic theses that determine both *Empire* and much of the theoretical discourse of contemporary post-socialist anti-capitalism.

TRONTI'S COPERNICAN REVOLUTION

The source for this turn to an explicitly and systemically antagonistic brand of Marxism is twofold. Historically speaking, it was born of the resurgence—outside of the direct sway of the PCI (Italian Communist Party) and official trade unions—of fierce workers' struggles in the late 1950s and throughout the 1960s, where what was at stake was no longer the participation in the nationalist and productivist agenda of progress and negotiation, but rather the unilateral demand for the *immediate* satisfaction of workers' needs outside of any rationale that would see these needs as predicated upon the buoyancy of the economy, the continuation of high levels of investment, and a general increase in production and profitability. Theoretically speaking, this wave of openly 'egotistical' struggles, marked by the refusal of any socialist idolatry of work as the essence of the human as well as by an utter disdain for the political impetus behind economic plans, was eminently registered in Mario Tronti's epoch-making *Ope-*

---

9. Negri, *Marx Beyond Marx*, p. 168.

rai e capitale [*Workers and Capital*].[10] This work, together with the productions of some of Tronti's comrades in the journal *Quaderni Rossi*, tried to operate a radical reversal of the theoretical standpoint that regarded labour-power as a *factor* within the cycles of production and their political rationalization. This was a factor that could at best delegate political command over itself to the party as class representative, but which, until the attainment of the receding threshold of communism, would remain fettered by the demanding discipline of the essentially capitalist relations obtaining in the factory and beyond.

Against this ideology of productivism, economic planning and worker sacrifice, Tronti attempted to translate the antagonistic demands for appropriation that had marked ten years of workers' struggles into an adequate theoretical framework. Contrary to the view whereby it was possible interminably to engage capital in reformist political mediations safeguarding the livelihood (if not the desires) of the working class, Tronti argued for the illusory character of this position, on the basis of the following thesis, which becomes more persuasive by the day: The political history of capital is *the history of the successive attempts of the capitalist class to emancipate itself from the working class*. The strategic ambivalence of the working class as a subject of exploitation was framed by Tronti in the following characteristically lapidary lines:

> The working class *does* what it *is*. But it is, at one and the same time, the *articulation* of capital, and its *dissolution*. Capitalist power seeks to use the workers' antagonistic will-to-struggle as a motor of its own development. The workers' party must take this same real mediation by the workers of capital's interests and organize it in an antagonistic form, as the tactical terrain of struggle and as a strategic potential for destruction.[11]

What we have here is neither an organic dialectic nor a Manichean theory of pure antagonism. Rather we are introduced to the idea that capital is concerned with a *dialectical use of antagonism*, whose ultimate if utopian horizon is the withering away of the working class and the untrammelled self-valorization of capi-

---

10. See Mario Tronti, *Operai e capitale*, 3rd ed., Roma, DeriveApprodi, 2006.
11. Mario Tronti, 'The Strategy of Refusal', in S. Lotringer and C. Marazzi (eds.), *Autonomia: Post-Political Politics*, New York, Semiotext(e), 1980, p. 29.

tal; whilst the working class and its political vanguard aim at an *antagonistic use of antagonism*, which refuses precisely the capitalization of antagonism whereby, for example, the flight from the factory is turned into an opportunity for profitable technological leaps and the exploitation of a de-unionized 'flexible' work force. In Harry Cleaver's useful gloss, this means that 'capital seeks to incorporate the working class within itself simply as labour-power, whereas the working class affirms itself as an independent class-for-itself only through struggles which rupture capital's self-reproduction'.[12] Communist politics is thus aimed at exploiting the inner tensions of a capitalism whose 'strength' lies in the 'production of ever-renewed antagonism',[13] and which depends on 'breaking the autonomy of labour-power without destroying its antagonistic character'.[14]

In a sense, the exasperation of capital's bid for freedom, which became more obvious in 1960s in the transformation of the organic composition of capital (ratio of constant to variable capital, specifically involving an increase in controllable technologies and the marginalization of uncontrollable workers for the sake of increased productivity) did nothing but reveal that process, indicated by Marx in the *Results of the Immediate Process of Production*, whereby the working class (*qua* living labour) confronts the seemingly monolithic character of capital's command over the production process.[15] Here then lies the vampirism of capital, whose only fluidity is offered by the process of absorption of living labour. As Bruno Maffi, editor of the *Results* in Italian, noted: 'Capital is truly *capital* only if it becomes "value in process"; only if, *within* the process of production, the magic touch of human labour transforms it from a constant to a variable magnitude'.[16]

This dual phenomenology of the production process, split between the immediate point of view of production and the point of view of capital's self-valorization, is precisely the object of Tronti's attempt at forcing a political assumption of this antag-

---

12. Harry Cleaver, *Reading Capital Politically*, 2nd ed., London, AK Press, 2000, p. 66.

13. Mario Tronti, *Il tempo della politica*, Roma, Editori Riuniti, 1980, p. 58.

14. Tronti, *Operai e capitale*, p. 217.

15. Karl Marx, *Capital, Volume 1: A Critique of Political Economy*, trans. Ben Fowkes, London, Penguin, 1992, pp. 987-8.

16. Karl Marx, *Il Capitale, Libro I, Capitolo VI Inedito*, Bruno Maffi (ed.), Milan, Etas, 2002, p. xi.

onism, in the here and now, which would not subordinate itself to economic rationalization (which is always the prelude to capital's emancipation from the working class). By facing the totality of the conditions of labour as capital, alongside the increasingly intimate bond between these conditions and a practice of command and discipline (such that exploitation is sedimented by and articulated through objective technologies of discipline in production), we can, according to Tronti, begin to project the political constitution, through antagonism, of an explicitly militant and anti-systemic working class. On the terrain of the command over production, what serves as a structural or phenomenological antagonism must be assumed, doubled and reinforced (to the point of crisis) by a political antagonism that directly targets the capitalistic process of self-valorization, and tends towards a self-valorization of the working class, which is to say, towards a destabilization and de-structuring of capitalist command. The entire issue, both strategically and tactically (and the deep cause of numerous splits on the Italian left), concerned the means of moving from certain practices of autonomy that characterized workers' struggles to the political formation of what Tronti refers to as a *class against capital*. From insurrection to organization, and back again.

This is Tronti's 'Copernican revolution', whereby 'the economic laws of the movement of capitalist society must be newly discovered as the political laws of the movement of the working class' and '*bent* with subjective force of organization brutally to serve the objective revolutionary needs of antagonism and struggle'.[17] Capital, through this openly political torsion, becomes a function of the working class, in a situation wherein politics 'precedes' science. As Cristina Corradi has duly noted in her recent history of Italian Marxism, if we wish to stick with the scientific analogy, this Copernican revolution is really a 'post-Copernican', or Einsteinian one. Tronti's vision of a new politicized antagonistic science of capital is not that of a 'general methodology and universal science' but of a *partial, subjective, unilateral science, in the ambit of a system marked by a high degree of indeterminacy*. The Marxist inquiry is compared to the discovery of non-Euclidean geometries, just as the spirit of the October revolution is argued to have an affinity with the break represented by Ein-

---

17. Tronti, *Operai e capitale*, p. 224.

stein's theory of relativity'.[18] This idea of a partisan science of capital, which dominates Tronti's work and is also present, in a different guise, in Negri, has a number of significant consequences, two of which I want to mention. First, it entails that there is no scientific theory from which one could simply deduce political action. Rather, theory as an attempt to grasp the objective tendencies of accumulation is always in a relation of disjunctive synthesis to politics as the 'global refusal of objectivity', the attempt to vanquish the tendency. In other words: 'Theory is anticipation. Politics is intervention'.[19] Furthermore, it means that the link between politics as a science of intervention and Marxism as a science of anticipation must always be conquered in and against changing conjunctures: 'Science as struggle is an ephemeral knowledge. It lasts as long as it's useful…. This is a happy condition of thinking: when you know that there is one part, and one part only, of the world that asks you a question. A state of exception in which thinking is the force that decides'.[20] And contrary to a facile determinism, 'to predict the development of capital does not mean subjecting oneself to its iron laws: it means forcing it to take a path, waiting for it at some juncture with weapons stronger than iron, attacking and breaking it at that point'.[21] Crucially this link between tendency and initiative in 'brief political moments' can mean that certain opportunities for ambushing capital can be irretrievably lost, that defeat is a real possibility. As Tronti warns in *Operai e capitale*, 'we don't have much time'.[22]

Tronti's work does not simply represent a voluntaristic adjunct to the critique of political economy, but wishes to recast capitalist society and capitalist domination as a *reactive formation*, a character recognized by Marx himself in his accounts of the theft of workers' knowledge and ensuing structural adjustments in the process of production. As Marx once quipped, capital (with all its technological prostheses) chases strikes. The key axiom here, which proved a huge influence on Negri's work throughout the 70s, and which remains embedded in the latest analyses of

---

18. Cristina Corradi, *Storia dei marxismi in Italia*, Roma, manifestolibri, 2005, p. 169.
19. Corradi, *Storia dei marxismi in Italia*, p. 258.
20. Mario Tronti, *Cenni di castella*, Fiesole, Cadmo, 2001, p. 19.
21. Tronti, *Il tempo della politica*, p. 64, Tronti, *Operai e capitale*, p. 17.
22. Tronti, *Operai e capitale*, p. 21.

the 'multitude' is the following: there is a primacy of resistance over exploitation and domination. The corollary of this axiom is that 'capital is a consequence of worker's labour'. In Tronti's own words: 'it is the specific moments of the class struggle which have determined every technological change in the mechanisms of industry'.[23] Contrary to Tronti's later stance, which would see the possibility, heralded by the 'political centrality of the working class' of a communist use of 'the provisional autonomy of state manoeuvres from capitalist interest'[24] (echoing the PCI's view of itself as a superior organizer of capitalist production), his writings of the early and mid-1960s exude a combative *irresponsibility* on the part of the working class within a society riven by antagonism: 'It is not up to the workers to resolve the conjunctures of capitalism. Let the bosses do it, on their own. It is their system: let them sort it out. It is here that a strategy of the total refusal of capitalist society must find the positive tactical forms for the most effective aggression against the concrete power of capitalists'.[25] Against the neutrality of technology, its manipulation and 'evolution', and against any productivist compact between big government, big business, big unions and a big party, this position argues for the use of the political antagonism of labour and capital as a prism for comprehending the dynamics of social transformations in terms of the subjection and absorption of living labour by dead capital, foregrounding the subjectivity of the working class, which is both the presupposition and the principal threat to capitalist reproduction.

It is on this basis that Tronti articulates the paradoxical situation of workers labouring under capitalist command: 'the only thing which does not come from the workers is, precisely, [the conditions of] labour'.[26] That is, it is the overtly political framework of command, discipline and rationalization of the labour process that serves to shackle living labour to the demands of capital, such that the 'ontological' primacy and ineluctability of living labour is subjected to a thoroughgoing instrumentalization. As Marx himself had acerbically indicated: 'It is not the worker who buys the means of production and subsistence, but

---

23. Tronti, 'The Strategy of Refusal', p. 30.
24. Tronti, *Il tempo della politica*, p. 64.
25. Tronti, *Operai e capitale*, p. 98.
26. Tronti, 'The Strategy of Refusal', pp. 30-1.

the means of subsistence that buy the worker to incorporate him into the means of production'.[27] But for Tronti, Negri and their comrades, in the phase of 'high' workerism, these mechanisms of coercion that situate the bearer of labour-power within the system of production, circulation and distribution mask the very real *dependency* of capital, which cannot be simply dispelled by means of changes in the organic composition of capital. Capitalism is both thoroughly dependent upon the capacity, relative docility and availability of the working class and constantly dreams of (often brutally destructive) ways of escaping this dependency; of escaping the moment of labour in the cycles of accumulation. As Tronti writes, 'Exploitation is born, historically, from the necessity for capital to escape from its *de facto* subordination to the class of worker-producers'.[28]

Thus, it can be argued that capital is in a double bind, which demands from it both a ruthless command and minimization of workers' demands (or at least of any of those demands that would interfere with capitalist valorization) *and* a capacity to absorb not simply living labour in terms of the physical expenditure of the worker, but a whole host of skills, knowledges and capacities for cooperation that are inseparable from workers' struggles for an emancipation *from* and not *of* work. The problem of capitalist command becomes that of a parasitic capture of the political vitality of the working class joined to a neutralization of its deeply threatening nature. This is where Tronti points to the role of 'organic forms of political dictatorship' in the history of capitalism, and we may consider today the twin phenomena of the *grand enfermement* of the American 'underclass' and the punitive and selective measures aimed at migrants in Europe and elsewhere in this light.[29] The paramount function, within social conflict, of

---

27. Marx, *Capital, Volume 1: A Critique of Political Economy*, p. 1004.

28. Tronti, 'The Strategy of Refusal', p. 30.

29. The continuing vitality of the partisan methodology of workerism—linking the study of class composition, the primacy of struggle and the forms of capitalist dictatorship—is evident in the work of a generation of researchers who have combined its prescriptions with the tools of other radical theoretical traditions (from the Foucauldian and Deleuzian study of societies of discipline and societies of control, to notions of subjectivation originating in subaltern studies and postcolonial theory). Alessandro De Giorgi's studies of postfordist regimes of penality (Alessandro De Giorgi, *Zero Tolleranza. Strategie e pratiche della società di controllo*, Rome, DeriveApprodi, 2000, Alessandro De Giorgi, *Il governo dell'eccedenza. Postfordismo e controllo della moltitudine*, Verona, Om-

the state *of* capitalism means that the antagonism at the heart of the process of production can only manifest itself as an attack on the state, what Negri would call a destabilization and a de-structuring. Tronti's *Operai e capitale* outlines the tendency towards the ever more explicit face-off between two separate but reciprocal processes of subjectivation: the subject of capitalist command and the subject of communist insurrection. Here Tronti introduces the specific *political difference* of labour and capital: the first does not need institutions, but only organization, while the second must be institutionally articulated. As he writes:

> From the very beginning, the proletariat is nothing more than the immediate *political interest* in the abolition of every aspect of the existing order. As far as its internal development is concerned, it has no need of 'institutions' in order to bring to life what it is, since what it is is nothing other than the *life-force* of that immediate destruction. It doesn't need *institutions*, but it does need *organization*. ... *The concept of the revolution and the reality of the working class are one and the same.*[30]

Against a social-democratic politics of mediation, Tronti argues that the strategic setbacks of the working class movement have always been based on seeking to transfer the model of the bourgeois revolution to the communist revolution—to wit, of imagining a slow takeover of economic power, followed by the reversal of political control.[31] In other words, the perpetual delay

---

bre Corte, 2002) and Sandro Mezzadra's *Diritto di fuga*, with its thesis on the 'autonomy of migration', are of great significance in this regard (see Sandro Mezzadra, *Diritto di fuga. Migrazioni, cittadinanza, globalizzazione*, 2nd ed., Verona, Ombre Corte, 2006). For an insightful post-workerist attempt to think struggle, discipline and control in terms of the transformations and uses of money, see the collective volume *La moneta nell'Impero*, especially Andrea Fumagalli's 'Moneta e potere: controllo e disciplina sociale' (see Andrea Fumagalli, Christian Marazzi and Adelino Zanini, *La moneta nell'Impero*, Verona, Ombre Corte, 2002). It is on the subjective side, which is to say vis-à-vis the articulation between class and organization, that these texts show the contemporary difficulties facing a workerist legacy. In this respect, the concept of 'multitude' seems to serve more as a place-holder than a solution when it comes to the present impasses of a politics of working class insurrection.

30. Tronti, 'The Strategy of Refusal', p. 34.

31. In his later, more melancholic reflections on the closure of twentieth-century political subjectivity, Tronti will note that it is the very illusion of social-democracy that it can subsist without the fire of insurrection: 'No reformist practice can advance if it is not accompanied, fuelled, and given substance by

of a full assumption of antagonism and autonomy on the part of working-class movements has meant that:

> Basically, all the communist movement has done has been to break and overturn, in some aspects of its practice, the social democratic logic of what has been its own theory ... here we see the working class articulation of political development: at first as an initiative that is positive for the functioning of the system, an initiative that only needs to be organized via institutions; in the second instance, as a 'No', a refusal to manage the mechanism of society as it stands, merely to improve it—a 'No' which is repressed by pure violence. This is the difference of content which can exist—even within one and the same set of working class demands—between *trade union demands* and *political refusal*.[32]

FANTASY WEARS BOOTS

Whilst Tronti—convinced that the workers' movement could only be articulated through a mass party—returned to the PCI and tried to formulate the idea of an 'autonomy of the political' as a way of achieving working class hegemony over economic planning and rationalization (as part of a theoretical shift skillfully tracked by Matteo Mandarini in this same issue), Negri's entire political and theoretical development is founded on the non-dialectical intensification of antagonism. The aim was to find an insurrectional and organizational outlet for Tronti's exhortation: 'As a matter of urgency we must get hold of, and start circulating, a photograph of the worker-proletariat that shows him as he really is—"proud and menacing"'.[33] Negri's turn to an expanded reproduction of antagonism throughout the social sphere, beyond the factory and the mass party, depended once again on a certain assessment of the tendency at work within late capitalism, a tendency characterized by an ever-increasing excercise of command, crisis and control on the side of capital, aimed at the subjection of workers, the decomposition of any possible form of class unity and an extraction of surplus-value that tries to eman-

---

a thinking of revolution', see Mario Tronti, *La politica al tramonto*, Torino, Einaudi, 1998, p. 52.

32. Tronti, 'The Strategy of Refusal', p. 34.
33. Tronti, 'The Strategy of Refusal', p. 34.

cipate itself from any dialectic or negotiation with the bearers of labour-power.

In this phenomenon of tendency—which included the blackmail of austerity policies, the Cold War's nuclear emergencies, and the ever increasing role of monetary policies after the oil crisis of 1973—Negri registers an increasing violence and irrationality on the part of capital. This violence ultimately lies in trying to maintain the *measure* and *command* of salary relations in a situation where social cooperation and technological advance are at such a level that the continuation of exploitative relations becomes ever more nonsensical. The 'crisis politics' and 'strategy of tension' that characterized the Italian state, but also the violent class decomposition that marked the onslaught against organized labour by Thatcherism and Reaganism, making way for the present neoliberal regime of flexibility, are emblems of the necessary vertical force required to reproduce capitalist social relations. As Negri remarks:

> My denunciation is not therefore directed against the normality of violence, but against the fact that in the enterprise form of capitalist domination, violence has lost all intrinsic, 'natural' rationale ('naturalness' being always a product of historic forces), and all relation with any project that could be deemed progressive. If anything, the enterprise form of violence is precisely the opposite: it is an irrational form within which exchange value is imposed on social relations in which the conditions of the exchange relation no longer exist. It is the intelligent form of this irrationality, simultaneously desperate in its content and rational in its effectiveness.[34]

In these passages, albeit in a far less morbid and claustrophobic vein, Negri anticipates the analysis of post-historical character of state violence forward by Debord in his *Commentaries on the Society of the Spectacle*, and later seconded by Giorgio Agamben, who writes of how in 1970s Italy 'the governments and servants of the entire world had observed then with attentive participation ... the way that a well-aimed politics of terrorism could possibly function as the mechanism of relegitimation of a discredited system'.[35] But for Negri the collapse of the dialectics of value and

---

34. Negri, *Revolution Retrieved*, p. 131.
35. Giorgio Agamben, *Means Without End: Notes on Politics*, trans. Vincen-

measure still has its source in the subjective pressure of antagonism, and indeed of constituent power. This means that the capitalist use of crisis and emergency, or rather the emergence of a 'crisis state' cannot be metaphysically and trans-historically sublimated into a view, such as Agamben's, whereby *'the state of exception is the rule'* and 'naked life ... is today abandoned to a kind of violence that is all the more effective for being anonymous and quotidian'.[36] Contra Agamben, for Negri, then and now (as his critiques of the thesis of bare life make evident), this violence is always a determinately *capitalist* violence, that is to say a violence that *reacts* against a primary resistance, or better a prior antagonistic production of subjectivity.

Thus the tendency to an integral *socialization* of capitalism (following the *Grundrisse*, the 'bible' of *operaismo*), spreading far beyond the factory gates and encompassing all facets of social reproduction within the extraction of surplus value, comes into conflict with the endurance, enforced by exquisitely political means, of the measurability of production in the form of the wage. Arguing from the loss of any proportionality or translatability between a production now entirely socialized (the thesis of real subsumption) and its measure in labour-power or wage, Negri, beginning in the 1970s, identifies the tendency as the site of a *communist transition*. This transition however does not take the form of a plan or programme, but of an outright refusal of capitalist command and a consequent reappropriation—on the basis of an analysis of class composition, that is to say of the power-relations and differentiations within the working class itself—of workers' experience and productivity. The self-valorization of capital through command is thus confronted by the self-valorization of the working class via practices of autonomy aimed at destabilising and de-structuring of the political conditions for the perpetuation of capitalism. The programme is thus that of 'the direct social appropriation of produced social wealth'.

It is here that the concrete practices of the movements gathered under the banner of *Autonomia organizzata*—agitating in Rome, Padua, Milan and other urban areas in the 70s, and supported by publications such as *Rosso*—find their theoretical le-

---

zo Binetti and Cesare Casarino, Minneapolis, University of Minnesota Press, 2000, p. 127.

36. Agamben, *Means Without End*, p. 113.

gitimacy. The practice of mass illegality (unilateral reduction of bills, house occupations, and so on), sabotage and violent assertions of the material reality of worker independence, all of which characterized the 'autonomist' movement in the 1970s, are thus conceptualized as an attempt to force the structural antagonism and its tendency towards an ever-greater arbitrariness of command. This strategy, not just of refusal but of the conquest of metropolitan 'red bases' and the irrecuperable intensification of antagonism, was aimed at preparing a generalized insurrectionary situation. The assumption of autonomy was thought to function directly as means of destabilising and destructuring, recomposing class unity and countering the neutralization of resistance that the capitalist state effects through means both punitive (repressions and redundancies) and programmatic (the decomposition of a factory-based working class and creation of a precarious and flexible class of 'immaterial' workers: a situation that backfired in 1977, when the micropolitical strategies of the crisis-State—dispersion of workers, flexibilization—led to mass uprisings of unemployed and often highly educated urban youth).

This insurrectionary program is based on an analysis of a twofold tendency. On the one hand, we have the increasingly brutal attempt on the part of capital to emancipate itself from workers and workers' struggles, its 'dream of self-sufficiency'. On the other, we are presented with the increasing socialization of value, such that processes of production and reproduction, as well as circulation and distribution, become increasingly integrated and less and less linked to the mediating space of the factory and the official working class movement. The antagonism is therefore posited as an extreme contest between, on the one side, a capital hell-bent on the absoluteness of its own command and the fragmentation of any class initiative; and, on the other, a class of social workers (*operai sociali*, the mutant descendants of the Fordist mass worker) striving to attain a direct appropriation of the social production that finds its source in their own living labour as well as in their everyday practices and desires (chiefly in the domain of a consumption that is integrally 'put to work'). The subjectivation, singularization and socialization of living labour is thus the aim of a movement that seeks to force the separation from capitalist command.

But it is a subjectivation that, as we move into the 1970s and the decomposition of the factory, is obliged to spread itself across the entire social field. This is where the concept of class composition and the analysis of power-relations is of such importance, as without it only an entirely indeterminate dualism of class against state—ripe for a vanguardist and terrorist takeover *à la* Red Brigades—can take place. Here is where we encounter the fundamental non-homogeneity of class composition, the emergence of a *disseminated* figure of the worker and the need to generate new organizations of class struggle on a new terrain. In this context, the politicization of marginal labour power into working class is never given (in the factory, in the 'movement') but must be conquered explicitly. This is where the notion of the 'refusal of work'—to be understood as the refusal of the reproduction of capitalist wage-relations for the sake of an emancipation of social production, or of what Negri calls the 'force of invention'—takes root and acquires a pivotal role. Refusal of work, articulated outside the factory, is aimed both at class unity (crystallization of a new class composition beyond the factory) and geared for the project of destroying capitalist relations by the unconditional demand for a *right to income*, a *political wage* entirely detached if not wholly destructive of the conditions for the reproduction of capitalist cycles of profit and investment (this proposal returns in a slightly different guise in both *Empire* and *Multitude*).

Ultimately, the very terms of the antagonism, of the 'method of tendency' espoused by Negri, do demand the confrontation—determined by the particularities of class composition, organic composition and capital's strategies of restructuring and command, but neither mediated or dialogical—between the violence of a command that tries to maintain the wage-relation and the measure of labour-power, on the one hand, and the creative violence of a self-valorising working class, on the other. We could thus say that both the force and the shortcomings of Negri's position lie in his determination to sap any possibility of institutional compromise, and in his insistence in addressing the question of *power* in its two senses of power over the state (of capital) and of power-relations within classes themselves (class composition). To use the Spinozist distinction so dear to him, we have here the face-off between the *potentia* of the working class and the *potestas* of a State dominated by the logic of the enterprise, the firm. If the face-off cannot be avoided, whatever its forms, it is because the very analysis of tendency means that a counter-autonomy or counter-self-valorization—briefly, insurrection—is the only countervailing force against the violence of capitalist command over the socialization of production. As Negri says, in discord with some of his later pronouncements about the exodus of the multitude: 'The *jouissance* that the working class seeks is the *jouissance* of power, not the tickle of illusions'. This theme returns in other texts from his 1970s Feltrinelli pamphlets, confiscated and immolated by the very state whose violence they dissected: 'Fantasy wears boots, desire is violent, invention is organized'. And further: 'The Party is the army that defends the borders of proletarian independence'. But this counter-violence against the state, which is the violence of a sabotage aimed both at the defence of worker's needs and experiences, and at the destruction of capitalist relations, was forced by its objective weakness into a strategy that could easily be portrayed as one provocation; a strategy which, at least in the Italian case, proved that, alas, in Negri's own words: 'Crisis is a risk taken by the working class and the proletariat. Communism is not inevitable'.[37]

Where the insurrectionary *élan* of *operaismo* for a time promised a refusal and a separation from a position of strength (in the conviction that the primacy of resistance heralded the eventual obsolescence of capitalist command), the current conjuncture— witness the 'post-workerist' writings of Marazzi, De Giorgi, Fumagalli, Vercellone and several others—leads to an inevitable preoccupation not so much with separation or autonomy, as with the identification of subjective and material levers to disarticulate forms of command that have grown more recondite and redoutable since the 1970s. The challenge today is to think an antagonism whose autonomy would not entail a doomed attempt at separation, an antagonism that would not be entirely detached from the conditions of production and reproduction of contemporary capitalism. The mere positing of a duality, say between Empire and multitude, without the conflictual *composition* that can provide this duality with a certain degree of determinateness, can arguably be seen to generate a seemingly heroic, but ultimately ineffectual horizon for theoretical analysis and political militancy. In political-historical figures such as those of the 'im-

---

37. Antonio Negri, *Books for Burning*, Timothy S. Murphy (ed.), London, Verso, 2005, pp. 39, 260, 276, 280 (translation modified).

material labourer', a certain post-workerism seems to glimpse not just the end of the measured dialectic of capital and labour, but the overcoming of the need politically to confront the violence of capitalist command. Negri himself sees his work as leading to the 'theoretical observation that the social transformation of class relations is definitively over. Today, *against* capital, rises up the social figure of immaterial labour'.[38]

In this regard, any work that seeks to reinject the workerist method of antagonism into the current composition of social relations, into the uneven and combined development of capitalist command and political struggles, will be obliged to tackle two questions: How do we confront a situation in which capitalism's vicious rounds of accumulation by dispossession point to its continued and virulent, if contradictory, desire to emancipate itself from the working class, if not from humanity as a whole? And what does it mean to revive or prolong the methodologies and political gestures of workerism and autonomy at a time when—in many of the core capitalist economies that were always the privileged terrain of workerism—we are confronted by 'a depoliticization of society that reinforces the power of dominant forces'?[39]

---

38. Negri, *Books for Burning*, p. xlix (translation modified).

39. Mario Tronti, *Con le spalle al futuro. Per un altro dizionario politico*, Roma, Editori Riuniti, 1992, p. 13.

10 NATURAL-HISTORICAL DIAGRAMS: THE 'NEW GLOBAL' MOVEMENT AND THE BIOLOGICAL INVARIANT

Paolo Virno

1. ALWAYS ALREADY JUST NOW

The content of the global movement which ever since the Seattle revolt has occupied (and redefined) the public sphere is nothing less than human nature. The latter constitutes both the arena of struggle and its stake. The arena of struggle: the movement is rooted in the epoch in which the capitalist organization of work takes on as its raw material the differential traits of the species (verbal thought, the transindividual character of the mind, neoteny, the lack of specialized instincts, etc.). That is, it is rooted in the epoch in which human praxis is applied in the most direct and systematic way to the ensemble of requirements that make praxis human. The stake: those who struggle against the man-traps placed on the paths of migrants or against copyright on scientific research raise the question of the different socio-political expression that could be given, here and now, to certain biological prerogatives of *Homo sapiens*.

We are therefore dealing with a historically determinate subversive movement, which has emerged in quite peculiar, or rather unrepeatable, circumstances, but which is intimately concerned with that which has remained unaltered from the Cro-Magnons onwards. Its distinguishing trait is the extremely tight entanglement between 'always already' (human nature) and 'just now' (the bio-linguistic capitalism which has followed Fordism and Taylorism). This entanglement cannot fail to fuel some Rousseauian conceptual muddles: the temptation to deduce a socio-political ideal from the biological constitution of the human animal seems irrepressible, as does the idea of a naturalist corrective to the distortions produced by an irascible history. Think of the political Chomsky, for whom the crucial point is to constantly reaffirm some innate capabilities of our species (for example, the 'creativity of language'), against the claims, unjust because *unnatural*, of this or that system of power. To my mind, there is both truth and falsehood in the 'Chomskyianism' that pervades the common sense of the movement. Truth: it is absolutely realistic to hold that the biological invariant has today become a fulcrum of social conflicts, in other words that immutable metahistory surges up at the centre of the most up-to-date labour and communicative processes. Falsehood: the biological invariant becomes the raw material of social praxis only because the capitalist relation of production mobilizes to its advantage, in a historically unprecedented way, the specie-specific prerogatives of *Homo sapiens*. The undeniable preeminence of the meta-historical plane entirely depends on a contingent state of affairs.

To clarify the link between global movement and human nature it is necessary to tackle, be it tangentially, some tricky problems. First (§2), an apodictic thesis: how and why is human nature, far from being only the condition of possibility of historical praxis, also at times its manifest content and operational field. Second (§3), a synoptic definition, itself also apodictic, of those phylogenetic constants which are simultaneously the condition of possibility and the manifest content of historical praxis. On the basis of these premises,[1] the real discussion begins. It consists (§§4-5) in confronting the rather different ways in which the background, that is human nature, comes to the foreground, in the guise of an empirical phenomenon, in traditional societies and in contemporary capitalism. This crucial difference helps us to better understand the specific weight which the political action of the global movement carries, or could carry (§6).

2. MAPS OF HUMAN NATURE

The decisive question is broadly the following: can human beings *experience* human nature? Note that experiencing something, for

---

1. Which are fully argued in Paolo Virno, *Quando il verbo si fa carne. Linguaggio e natura umana*, Turin, Bollati Boringhieri, 2003.

instance an object or an event, does not at all mean representing it with some degree of scientific precision. Rather, it means perceiving it in its phenomenal manifestness, being emotionally involved, reacting to it with praxis and discourse. If that is so, our case immediately confronts us with a difficulty: the expression 'human nature' effectively denotes the ensemble of innate dispositions that guarantee the very possibility of perceiving phenomena, to be emotionally involved, to act and discourse. Accordingly, the decisive question takes on a paradoxical air: is it possible to experience, in the full sense of the term, that which constitutes the presupposition of experience in general?

The answer depends on the way in which we conceive of eternity in time. Make no mistake: by 'eternal' I simply mean that which displays a high degree of invariance, not being subject to social and cultural transformations. In this mild acceptation, 'eternal', for instance, can be said of the language faculty. There are basically two ways, opposed to one another, of conceiving the eternal in time. The first, which I reject, can be loosely defined as 'transcendental'. Its point of honour lies in arguing that the invariant presuppositions of human nature, on which really experienced facts and states of affairs depend, never present themselves in turn as facts or states of affairs. The presuppositions remain confined in their recondite 'pre-'. That which grounds or permits all appearances does not itself appear. This approach rules out that human beings may experience human nature. The second way of considering the eternal in time can be defined, once again loosely, as 'natural-historical'. It consists in demonstrating that the conditions of possibility of human praxis possess a peculiar empirical counterpart. In other words, there are contingent phenomena which reproduce point-by-point the inner structure of the transcendental presupposition. Besides being their foundation, the 'eternal' exposes itself, as such, in such and such a given socio-political state of affairs. Not only does it *give* rise to the most varied events, but it also *takes* place in the flow of time, taking on an evental physiognomy. In other words, there are historical facts which show in filigree the conditions that make history itself possible. This second approach, which I share, implies that human beings can experience human nature.

I call *natural-historical diagrams* the socio-political states of affairs which display, in changing and rival forms, some salient features of anthropogenesis. The diagram is a sign that imitates the object to which it refers, meticulously reproducing its structure and the relation between its parts. Think of a map, a mathematical equation, a graph. However, the contingent historical fact, which offers the abridged image of a biological condition, is not a necessary condition of the latter, since its roots lie instead in a particular social and cultural conjuncture. The diagram faithfully reproduces the object that it stands for but, unlike an index, it is *not* caused by it. A geographical map is something other than the knock on the door which attests to the presence of a visitor.

Recall the question we formulated above: is it possible to experience, in the full sense of the term, that which constitutes the presupposition of experience in general? I can now reply: yes, if and when there are adequate phenomenal diagrams of this presupposition; yes, if and when a historical event offers the map or the equation of certain fundamental meta-historical constants. The diagrams of human nature institute an endless circularity between the transcendental and the empirical, the condition and the conditioned, the background and the foreground. To get an approximate idea of the diagram, consider this observation by Peirce on self-reflexive diagrams (I thank Tommaso Russo for having brought it to my attention): 'On a map of an island laid down upon the soil of that island there must, under all ordinary circumstances, be some position, some point, marked or not, that represents *qua* place on the map the very same point *qua* place on the island'.[2] The map is the diagram of a territory, part of which is constituted by the diagram of that territory, part of which... to infinity. The same happens, in effect, when you formulate a mental image of your own mind; accordingly, the image of the mind includes an image of the mind that includes an image... to infinity. Unlike the map discussed by Peirce, the diagrams of human nature are not scientific constructions or conventional signs; they are concrete phenomena, socio-political states of affairs, historical events. What's more, the paradoxical oscillation implied by these diagrams is not spatial but temporal. That is, it consists in the infinite circularity between 'just now' and 'always already' (experienced facts and conditions of possibility of experience); not in the circularity between part and whole, as in the case examined by Peirce.

---

2. Charles Sanders Peirce, *Collected Papers*, Charles Hartshorne and Paul Weiss (eds.), vol. 2, Cambridge, Harvard University Press, 1933, p. 230.

Natural history, in the particular sense I am giving to it here, meticulously collects the multiple socio-political *diagrams* of the biological invariant. Accordingly, it concerns itself with all the circumstances, rather different over the course of time, in which *anthropos*, working and speaking, retraces the salient stages of anthropogenesis. Natural history inventories the ways in which human beings *experience* human nature. Having the latter as its content, the global movement should be considered as an episode of natural history. It can rightfully be compared to the map of an island which is laid down on a precise point on the island itself.

3. THE POTENTIAL ANIMAL

Our theme is and remains the existence of natural-historical facts that have the value of diagrams (graphs, maps, etc.) of human nature. However, in order to discuss these diagrams with greater precision, it is necessary to establish some aspect of the object that they designate. What are we speaking about when we speak of species-specific prerogatives, of phylogenetic metahistory, of biological invariant? The following annotations are merely offered by way of orientation: nothing more than a road sign. Whoever doesn't share them, or thinks they fall short, can replace or complement them at will. The crucial point, I repeat, is not an exhaustive definition of that which in *Homo sapiens* remains unaltered from the Cro-Magnons onwards, but the ways in which the mutable course of history sometimes thematizes the 'eternal', even exhibiting it in concrete states of affairs.

The biological invariant that characterizes the existence of the human animal can be referred back to the philosophical concept of *dynamis*, power. From a temporal angle, power means *not-now*, untimeliness, a deficit of presence. And we should add that if there were no experience of the not-now, it would also be impossible to speak of a 'temporal angle'; it is precisely *dynamis* which, by dissolving the 'eternal present' of God and the non-human animal, gives rise to historical time. The potentiality of *Homo sapiens*: (a) is attested by the language faculty; (b) is inseparable from instinctual non-specialization; (c) originates in neoteny; (d) implies the absence of a univocal environment.

    a.   The language faculty is something other than the ensemble of historically determinate languages. It consists in a body's inborn capacity to emit articulate sounds, that is in the ensemble of biological and physiological requirements which make it possible to produce a statement. It is mistaken to treat the indeterminate power-to-speak as a proto-language spoken by the entire species (something like a universal Sanskrit). The faculty is a generic disposition, exempt from grammatical schemas, irreducible to a more or less extended congeries of possible statements. Language faculty means language *in potentia* or the power of language. And power is something non-actual and still undefined. Only the living being which is born aphasic has the language faculty. Or better: only the living being which lacks a repertoire of signals biunivocally correlated to the various configurations—harmful or beneficial—of the surrounding environment.

b. The language faculty confirms the instinctual poverty of the human animal, its incomplete character, the constant disorientation that sets it apart. Many philosophers argue that the language faculty is a highly specialized instinct. But they go on to add that it is a specialization for polyvalence and generalization, or even—which amounts to the same—an instinct to adopt behaviours that have not been preset. Now, to argue that the linguistic animal is supremely able in... doing without any particular ability is really to participate in the international festival of the sophism. Of course, the language faculty is an innate biological endowment. But not everything that is innate has the prerogatives of a univocal and detailed instinct. Despite being congenital, the capacity to speak is only *dynamis*, power. And power properly speaking, that is as distinguished from a well-defined catalogue of hypothetical performances, coincides with a state of indeterminacy and uncertainty. The animal that has language is a potential animal. But a potential animal is a *non-specialized* animal.

c. The phylogenetic basis of non-specialization is *neoteny*, that is the 'retention of formerly juvenile characteristics produced by retardation of somatic development'.[3] The

---

3. Stephen Jay Gould, *Ontogeny and Phylogeny*, Cambridge, Belknap Harvard,

generic and incomplete character of the human animal, the indecision that befalls it, in other words the *dynamis* which is consubstantial with it, are rooted in some of its organic and anatomical primitivisms, or, if you prefer, in its congenital incompleteness. *Homo sapiens* has 'a constitutively premature birth',[4] and precisely because of this it remains an 'indefinite animal'. Neoteny explains the instability of our species, as well as the related need for uninterrupted learning. A chronic infancy is matched by a chronic non-adaptation, to be mitigated in each case by social and cultural devices.

d. Biologically rooted in neoteny, the potentiality of the human animal has its objective correlate in the lack of a circumscribed and well-ordered environment in which to insert oneself with innate expertise once and for all. If an environment [*ambiente*] is the 'ensemble of conditions [...] which make it possible for a certain organism to survive thanks to its particular organization',[5] it goes without saying that a non-specialized organism is also an *out-of-place* [*disambientato*] organism. In such an organism perceptions are not harmoniously converted into univocal behaviours, but give rise to an overabundance of undifferentiated stimuli, which are not designed for a precise operational purpose. Lacking access to an ecological niche that would prolong its body like a prosthesis, the human animal exists in a state of insecurity even where there is no trace of specific dangers. We can certainly second the following assertion by Chomsky: 'the way we grow does not reflect properties of the physical environment but rather our essential nature'.[6] Provided we add, however, that 'our essential nature' is characterized in the first place by the absence of a determinate environment, and therefore by an enduring disorientation.

1977, p. 483.

4. See Marco Mazzeo, *Tatto e linguaggio*, Rome, Editori Riuniti, 2004; Adolf Portmann, *Aufbruch der Lebensforschung*, Zurich, Rhein Verlag, 1965.

5. Arnold Gehlen, *Philosophische Anthropologie und Handlungsleghre*, Frankfurt am Main, Klostermann, 1983, p. 112.

6. Noam Chomsky, *Language and the Problems of Knowledge: The Managua Lectures*, Cambridge, MIT Press, 1988, p. 151.

We said that the primary task of natural history consists in collecting the social and political events in which the human animal is put into direct relation with metahistory, that is with the unmodifiable constitution of its species. We call natural-historical those maximally contingent phenomena which offer plausible *diagrams* of an invariant human nature. The terse definitions we proposed above allow us to specify the overall argument. The questions that natural history must face up to are accordingly the following: In what socio-political situations does the non-biological specialization of *Homo sapiens* come to the fore? When and how does the generic language faculty, as distinct from historical languages, take on a leading role within a particular mode of production? What are the *diagrams* of neoteny? Which are the maps or graphs that will adequately portray the absence of a univocal environment?

The answer to these questions will shed light on an essential difference between traditional societies and contemporary capitalism. In other words, it will shed light on the unprecedented features of the historical situation in which the global movement of Genoa and Seattle finds itself operating.

4. CULTURAL APOCALYPSES

In traditional societies, including to some extent in classic industrial society, the potentiality (non-specialization, neoteny, etc.) of the human animal takes on the typical visibility of an empirical state of affairs only in an emergency situation, that is in the midst of a *crisis*. In ordinary circumstances, the species-specific biological background is instead concealed, or even contradicted, by the organization of work and solid communicative habits. What predominates thus is a robust discontinuity, or rather an antinomy, between 'nature' and 'culture'. Anyone who would object that this discontinuity is merely a mediocre cultural invention, to be chalked up to the bilious anthropocentrism of spiritualist philosophers, would be making his own life too easy, neglecting what is by far the most interesting task: to individuate the *biological* reasons for the enduring bifurcation between biology and society. A program to naturalize mind and language that would forsake a *naturalist* explanation of the divergence between 'culture' and 'nature', preferring to reduce the whole affair to a... clash of ideas, would be shamelessly incoherent.

Let's stick with well-known, even stereotypical formulations. We call potential the corporeal organism which, lacking its own environment, must wrestle with a vital context that is always partially undetermined, that is with a *world* in which a stream of perceptual stimuli is difficult to translate into an effective operational code. The world is not a particularly vast and varied environment, nor is it the class of all possible environments: rather, there is a world *only* where an environment is wanting. Social and political praxis provisionally compensates for this lack, building *pseudo-environments* within which omnilateral and indiscriminate stimuli are selected in view of advantageous actions. This praxis is thus opposed to its invariant and meta-historical invariant. Or rather, it attests it to the very extent that it tries to rectify it. If we wanted to turn once again to a concept drawn from Charles S. Peirce's semiotics, we could say that culture is a 'Sign by Contrast' of a species-specific instinctual deficit: a sign, that is, which denotes an object only by virtue of a polemical reaction to the object's qualities. Exposure to the world appears, above all and for the most part, as a necessary immunization from the world, that is as the assumption of repetitive and predictable behaviours. Non-specialization finds expression as a meticulous division of labour, as the hypertrophy of permanent roles and unilateral duties. Neoteny manifests itself as the ethico-political defense of neotenic indecision. As a device which is itself biological (that is, functional to the preservation of the species), culture aims at stabilizing the 'indefinite animal', to blunt or veil its disorientation [*disambientamento*], to reduce the *dynamis* that characterizes it to a circumscribed set of possible actions. Human *nature* is such as to often involve a contrast between its expressions and its premises.

On this background, which we've evoked with all the brevity of a musical refrain, there stands out a crucial point, which is instead redolent with nuances and subtleties. We've already alluded to it: in traditional societies, the biological invariant (language as distinct from languages, raw potentiality, non-specialization, neoteny, etc.) acquires a marked historical visibility when, and only when, a certain pseudo-environmental setup is subjected to violent transformative traction. This is the reason why natural history, if it is referred to traditional societies, coincides for the most part with the *story of a state of exception*. It scrupulously describes the situation in which a form of life loses any obviousness, becoming brittle and problematic. In other words, the situation in which cultural defenses misfire and one is forced to return for a moment to the 'primal scene' of the anthropogenetic process. It is in such conjunctures, and only in such conjunctures, that it is possible to garner vivid *diagrams* of human nature.

The collapse of a form of life, with the ensuing irruption of metahistory into the sphere of historical facts, is what Ernesto de Martino, one of the few original philosophers in twentieth-century Italy, called a 'cultural apocalypse'. With this term he designated the historically determinate occasion (economic disruption, sudden technological innovation, etc.) in which the very difference between language faculty and languages, inarticulate potentiality and well-structured grammars, world and environment, becomes visible to the naked eye, and is dramatically thematized. Among the multiple symptoms which for De Martino presage an 'apocalypse', there is one which possesses strategic importance. The undoing of a cultural constellation triggers, among other things, 'a semantic excess which is not reducible to determinate signifieds'.[7] We witness a progressive indetermination of speech: in other words, it becomes difficult to 'bend the signifier as possibility towards the signified as reality'[8]; untied from univocal referents, discourse takes on an 'obscure allusiveness', abiding within the chaotic domain of the power-to-say (a power-to-say that goes beyond any spoken word). Now, this 'semantic excess not reducible to determinate signifieds' is entirely equivalent to the language *faculty*. In the apocalyptic crisis of a form of life, the biologically innate faculty fully exhibits the gap which forever separates it from any given language. The primacy attained by an undulating power-to-say is matched by the abnormal fluidity of states of affairs and the growing uncertainty of behaviours. As de Martino writes: 'things refuse to remain within their domestic boundaries, shedding their quotidian operability, seemingly stripped of any memory of possible behav-

---

7. Ernesto De Martino, *La fine del mondo. Contributo all'analisi delle apocalissi culturali*, Torino, Einaudi, 2002 [1977], p. 89.

8. De Martino, *La fine del mondo. Contributo all'analisi delle apocalissi culturali*, p. 632.

iours'.[9] No longer selectively filtered by a complex of cultural habits, the world shows itself to be an amorphous and enigmatic context. The conflagration of the ethico-social order thus reveals two correlated aspects of invariant 'human nature': a language faculty distinct from languages and a world opposed to any (pseudo-)environment whatsoever.

This twofold revelation is nevertheless transitory and parenthetical. The ultimate outcome of the apocalypse or state of exception is the institution of new cultural niches, capable of concealing and blunting once again the biological 'always already', that is the inarticulate and chaotic *dynamis*. Rare and fleeting are the apocalyptic diagrams of human nature.

5. METAHISTORY AND SOCIAL PRAXIS

What was said in the preceding section only counts for traditional societies. Contemporary capitalism has radically modified the relation between unalterable phylogenetic prerogatives and historical praxis. Today, the prevailing forms of life do not veil but rather flaunt without any hesitation the differential traits of our species. In other words: the prevailing forms of life are a veritable inventory of *natural-historical diagrams*. The current organization of work does not allay the disorientation and instability of the human animal, but on the contrary takes them to their extreme and systematically valorizes them. Amorphous potentiality, that is the chronic persistence of infantile characteristics, does not menacingly flare in the midst of a crisis. Rather it permeates every aspect of the tritest routine. Far from dreading it, the society of generalized communication tries to profit from the 'semantic excess not reducible to determinate signifieds', thereby conferring the greatest relevance to the indeterminate language faculty. According to Hegel, philosophy's first task is to grasp its time with thought. This proverbial precept, akin to the chalk that grates against the blackboard for those who delight in studying the ahistorical mind of the isolated individual, needs to be updated in the following way: the paramount task of philosophy is to come to grips with the unprecedented superimposition of the eternal and the contingent, the biologically invari-

---

9. De Martino, *La fine del mondo. Contributo all'analisi delle apocalissi culturali*, p. 91.

ant and the socio-politically variable, which exclusively connotes the current epoch.

Let it be noted in passing that this superimposition accounts for the renewed prestige which for some decades now has been accorded to the notion of 'human nature'. It does not depend on the impressive tectonic shifts within the scientific community (Chomsky's pitiless critique against Skinner's *Verbal Behavior* or suchlike) but on an ensemble of social, economic and political conditions. To believe the opposite is yet another demonstration of culturalist idealism (of a very academic sort, to boot) on the part of those who nonetheless never fail to toot the horn of the program to naturalize mind and language. Human nature returns to the centre of attention not because we are finally dealing with biology rather than history, but because the biological prerogatives of the human animal have acquired undeniable historical relevance in the current productive process. That is, because we are confronted with a peculiar empirical manifestation of certain phylogenetic, which is to say metahistorical, constants that mark out the existence of *Homo sapiens*. If a *naturalist* explanation of the autonomy enjoyed by 'culture' in traditional societies is certainly welcome, so is a *historical* explanation of the centrality attained by (human) 'nature' in the midst of post-Fordist capitalism.

In our epoch, the object of natural history is not a state of emergency, but everyday administration. Instead of dwelling on the erosion of a cultural constellation, we now need to concern ourselves with the way it is fully in force. Natural history does not limit itself to scavenging through 'cultural apocalypses'. Instead it tightens its grip on the totality of contemporary events. Because biological metahistory no longer surges up at the edges of forms of life, where they get stuck and idle, but installs itself durably at their geometric centre, testifying to their regular functioning, all social phenomena can be rightfully considered as *natural-historical phenomena*.

The dearth of specialized instincts and the lack of a definite environment, which have been the same from the Cro-Magnons onwards, today appear as noteworthy economic resources. It is not difficult to register the patent correspondence between certain salient features of 'human nature' and the sociological categories which are best suited to the current situation. The biological non-specialization of *Homo sapiens* does not remain in the background, but gains maximal historical visibility as the universal *flexibility* of labour services. The only professional talent that really counts in post-Fordist production is the habit not to acquire lasting habits, that is the capacity to react promptly to the unusual. A univocal competence, modulated in its last detail, now constitutes an authentic handicap for those obliged to sell their labour-power. Again, neoteny, that is chronic infancy and the related need for continual training, translates, without any mediation, into the social rule of *permanent formation*. The shortcomings of the 'constitutively premature birth' are converted into productive virtues. What matters is not what is progressively learned (roles, techniques, etc.) but the display of the pure power to learn, which always exceeds its particular enactments. What's more, it is entirely evident that the *permanent precarity* of jobs, and even more the instability experienced by contemporary migrants, mirror in historically determinate ways the congenital lack of a uniform and predictable habitat. Precarity and nomadism lay bare at the social level the ceaseless and omnilateral pressure of a *world* that is never an environment. They induce a paradoxical familiarity with the stream of perceptual stimuli that do not allow themselves to be translated into univocal actions. This overabundance of undifferentiated solicitations is no longer true only in the final analysis, but it is true in the first analysis. It is not a disturbance to be dispelled, but the positive soil on which the current labour-process develops. Lastly, what is perhaps the most relevant and comprehensive point: inarticulate power, which is not reducible to a series of preset potential acts, acquires an extrinsic, or better pragmatic aspect in the commodity *labour-power*. This term effectively designates the ensemble of generically human psycho-physical faculties, which are precisely considered as mere *dynameis* that have yet to be applied. Today labour-power largely coincides with the language faculty. And the language faculty, qua labour-power, unmistakably shows its difference with regard to grammatically structured languages. Language faculty and labour-power lie on the border between biology and history—with the added proviso that in our epoch this very border has taken on precise historical lineaments.

To affirm that contemporary forms of life have as their emblem the language faculty, non-specialization, neoteny, loss of environment, does not at all entail arguing that they are unruly.

Far from it. Being conversant with omnilateral potentiality demands, as its inevitable counterpoint, the existence of far more detailed norms than the ones which are in force in a cultural pseudo-environment. Norms so detailed that they tend to hold for a single case, for a contingent and non-reproducible occasion. The flexibility of labour services implies the unlimited variability of rules, but also, for the brief period in which they remain in force, their tremendous rigidity. These are ad hoc rules, of the kind that prescribe in minute detail the way of carrying out a certain action and only that action. Precisely where it attains the greatest socio-political relevance, the innate language *faculty* mockingly manifests itself as a collection of elementary *signals*, suited to tackling a particular eventuality. The 'semantic excess which is not reducible to determinate signifieds' often flips over into a compulsive reliance on stereotyped formulae. In other words, it takes on the seemingly paradoxical guise of a semantic *deficit*. In both of its polarities, this oscillation depends on the sudden absence of stable and well-articulated *pseudo-environments*. No longer screened by a protective cultural niche, the world is experienced in all its indeterminacy and potentiality (semantic excess); but this patent indeterminacy, which each time is to be contained and diluted in different ways, provokes by way of reaction halting behaviours, obsessive tics, the drastic impoverishment of the *ars combinatoria*, the inflation in transient but harsh norms (semantic defect). Though on the one hand permanent formation and the precarity of employments guarantee the full exposure to the world, on the other they instigate the latter's recurrent reduction to a spectral or mawkish dollhouse. This accounts for the surprising marriage between generic language faculty and monotonous signals.

6. THE DEMAND FOR THE GOOD LIFE

Let's sum up. In traditional societies, the biological invariant was thrust to the fore when a form of life imploded and came undone; in contemporary capitalism, when everything functions regularly. Natural history, usually busy registering with seismographic precision crises and states of exception, is instead concerned today with the ordinary administration of the productive process. In our epoch, the biological requirements of *Homo sapi-*

ens (language faculty, non-specialization, neoteny, etc.) match up point-by-point with the most significant sociological categories (labour-power, flexibility, permanent formation, etc.).

Two phrases by Marx, taken from the *Economic and Philosophical Manuscripts of 1844*, are perfectly suited to the current situation. The first says: 'It can be seen how the history of *industry* and the *objective* existence of industry as it has developed is the *open* book of the essential powers of man, man's psychology present in tangible form. [...] A *psychology* for which this book [...] is closed can never become a *real* science'.[10] To paraphrase: today's industry—based on neoteny, the language faculty, potentiality—is the externalized, empirical, pragmatic image of the human psyche, of its invariant and metahistorical characteristics. Today's industry therefore constitutes the only dependable textbook for the philosophy of mind. Here is Marx's second phrase: 'The whole of history is a preparation, a development, for *"man"* to become the object of *sensuous* consciousness'.[11] Once we expunge the eschatological emphasis (history doesn't prepare anything, let it be clear) we can paraphrase as follows: in the epoch of flexibility and permanent formation, human nature now constitutes an almost perceptual evidence, as well as the immediate content of social praxis. In other words: every step they take, human beings directly experience that which constitutes the presupposition of experience in general.

The raw material of contemporary politics is to be found in natural-historical phenomena, that is in the contingent events in which the distinctive traits of our species come to light. I say *raw material*, not a canon or a guiding principle. All political orientations are effectively faced with a situation in which human praxis is systematically applied to the ensemble of the requirements that make praxis human. But they do so in the name of contrasting interests. The shared attention to the differential traits of the species gives rise to diametrically opposed aims, whose realization depends on the balance of forces they enjoy, not on their greater or lesser conformity to 'human nature'. It is in vain that Chomsky appeals to the unalterable biological endowment

---

10. Karl Marx, 'Economic and Philosophical Manuscripts of 1844', *Early Writings*, trans. Rodney Livingstone and Gregor Benton, London, Penguin, 1975, p. 354.

11. Marx, 'Economic and Philosophical Manuscripts of 1844', p. 355.

of *Homo sapiens* to rectify the inherent injustice of contemporary capitalism. Rather than constituting the platform and parameter for a possible emancipation, the congenital 'creativity of language' appears today as an ingredient in the despotic organization of work; or better, it appears as a profitable economic resource. To the extent that it attains an immediate empirical consistency, the biological invariant is part of the problem, and certainly not the solution.

The global movement is inscribed in this context. Not unlike its enemies, that is not unlike the politics that prolongs oppression, it too has considerable familiarity with the metahistory that incarnates itself in contingent states of affairs. But it strives to discern the various forms that could take on the manifestation of the 'always already' in the 'just now'. That the congenital *potentiality* of the human animal fully manifests itself at the socio-economic level is an irreversible matter of fact; but that in manifesting itself, this potentiality is obliged to take on the features of the commodity *labour-power* is by no means an inescapable fate. On the contrary, it is a momentary outcome, which one should intransigently struggle against. Likewise, it is not set in stone anywhere that the phenomenological correlate of the biological *non-specialization* of our species will continue to be, always and regardless, the servile *flexibility* flaunted by the contemporary labour-process. The socio-historical prominence of human nature does not attenuate but rather immeasurably enhances the specific impact (and the irreparable contingency) of political action.

The global movement is the conflictual interface of biolinguistic capitalism. It is precisely because (and not in spite) of this that it presents itself on the public stage as an *ethical* movement. The reason for this is easy to intuit. We have said that contemporary production implicates all the attitudes the distinguish our species: language, reflexivity, instinctual deficiency, etc. With a simplifying but not empty formula, we could even say that post-Fordism puts to work *life* as such. Now, if it is true that biolinguistic capitalism appropriates 'life', that is the set of specifically human faculties, it is pretty obvious that insubordination against it must focus on this same fact. The life that is included in flexible production is countered by the demand (which is pertinent because it is itself 'non-specialized') of a *good life*. And the search for the *good life* is the only concrete theme of the 'science of mores'. As numerous as its misfortunes may be, it is beyond doubt that the global movement has indicated the point of intersection between natural history and ethics.

Translated by Alberto Toscano

11 GIORGIO AGAMBEN'S FRANCISCAN ONTOLOGY

Lorenzo Chiesa

1. *HOMO SACER*: A POLITICAL HERO

Giorgio Agamben's critical analysis of biopolitics, a politics for which power 'confronts pure biological life without any mediation',[1] famously revolves around the notion of *homo sacer*. This notion is derived from an enigmatic figure of Roman law that, for Agamben, embodies both 'the originary "political" relation'[2] of the West and an 'essential function' in modern and contemporary politics.[3] In being the 'damned' [*sacer*] who may be killed and yet not sacrificed—the one who may be killed with impunity by any man, and yet not sacrificed to the gods—the sacred man represents a limit concept. In other words, the life of *homo sacer*, that is 'bare life', is excepted from both human jurisdiction—since in his case the application of the law on homicide is suspended—and divine law—since his killing cannot be regarded as a ritual purification.[4] However, this double exclusion of *homo sacer* is clearly at the same time a double capture of his bare life, absolutely exposed to violence and death, in the juridical order.[5] As Agamben writes, '*homo sacer* belongs to God in the

---

1. Giorgio Agamben, *Means Without End: Notes on Politics*, trans. Vincenzo Binetti and Cesare Casarino, Minneapolis, University of Minnesota Press, 2000, p. 41.
2. Giorgio Agamben, *Homo Sacer: Sovereign Power and Bare Life*, trans. Daniel Heller-Roazen, Stanford, Stanford University Press, 1998, p. 85.
3. Agamben, *Homo Sacer*, p. 8.
4. See especially Agamben, *Homo Sacer*, pp. 81-2.
5. 'Not simple natural life, but life exposed to death (bare life or sacred life) is the originary political element' (Agamben, *Homo Sacer*, p. 88).

form of unsacrificeability and is included in the community in the form of being able to be killed'.[6] For this reason, the structure of *sacratio* should be connected with that of sovereignty, or sovereign exception, on which the juridico-institutional foundations of modern and contemporary Western politics allegedly rely. Like *sacratio*, the sovereign exception founds itself on an inclusive exclusion. Indeed, the sovereign paradoxically lies, at the same time, 'outside and inside the juridical order'.[7] Just as in the case of *homo sacer*, the law applies to the sovereign in no longer applying to him: it is by means of its power of imposing death with impunity, and not through its ability to sanction a transgression, that the sovereign exception constitutes the originary form of law over life. From this Agamben can therefore conclude that:

> The sovereign and *homo sacer* present two symmetrical figures that have the same structure and are correlative: the sovereign is the one with respect to whom all men are potentially *homines sacri*, and *homo sacer* is the one with respect to whom all men act as sovereigns.[8]

At this stage, Agamben's logic of biopolitics as the logic of the symmetry between sovereign power and the sacredness of bare life should readily be understood in terms of its historico-ontological destiny. Although this theme is only hinted at in *Homo Sacer* (1995) and the volumes that follow it, Agamben resolutely maintains that biopolitics is inherently metaphysical. If on the one hand 'the inclusion of bare life in the political realm constitutes the original [...] nucleus of sovereign power' and 'biopolitics is at least as old as the sovereign exception',[9] on the other hand, this political nexus cannot be dissociated from the epochal situation of metaphysics. Here Agamben openly displays his Heideggerian legacy; bare life, that which in history is increasingly isolated by biopolitics as Western politics, must be strictly related to 'pure being', that which in history is increasingly isolated by Western metaphysics:[10]

> Politics [as biopolitics] appears as the truly fundamental

---

6. Agamben, *Homo Sacer*, p. 82.
7. Agamben, *Homo Sacer*, p. 15.
8. Agamben, *Homo Sacer*, p. 84.
9. Agamben, *Homo Sacer*, p. 6.
10. See Agamben, *Homo Sacer*, p. 182.

structure of Western metaphysics insofar as it occupies the threshold on which the relation between the living being and the *logos* is realized. In the 'politicization' of bare life—the metaphysical task *par excellence*—the humanity of living man is decided.[11]

Commentators have not as yet sufficiently emphasized how biopolitics is consequently nothing else than Agamben's name for metaphysics as nihilism. More specifically, while bare life remains for him the 'empty and indeterminate' concept of Western politics[12]—which is thus as such originally nihilistic—its forgetting goes together with the progressive coming to light of what it conceals. From this perspective, nihilism will therefore correspond to the modern and especially post-modern generalisation of the state of exception: 'The nihilism in which we are living is [...] nothing other than the coming to light of [...] the sovereign relation as such'.[13] In other words, nihilism reveals the paradox of the inclusive exclusion of bare life, *homo sacer, qua* foundation of sovereign power, as well as the fact that sovereign power cannot recognize itself for what it is. Beyond Foucault's biopolitical thesis according to which modernity is increasingly characterized by the way in which power directly captures life as such as its object, what interests Agamben the most is:

> The decisive fact that, together with the process by which exception everywhere becomes the rule, the realm of bare life—which is originally situated at the margins of the political order—gradually begins to coincide with the political realm.[14]

The political is thus reduced to the biopolitical: the original repression of the sovereign relation on which Western politics has always relied is now inextricably bound up with its return in the guise of a radical biopoliticisation of the political. Like nihilism, such a generalisation of the state of exception—the fact that, today, we are all virtually *homines sacri*—[15] is itself a profoundly ambiguous biopolitical phenomenon. Today's state of exception both radicalizes—qualitatively and quantitative-

---

11. Agamben, *Homo Sacer*, p. 8.
12. Agamben, *Homo Sacer*, p. 182.
13. Agamben, *Homo Sacer*, p. 51.
14. Agamben, *Homo Sacer*, p. 9. See also p. 38, p. 20.
15. See Agamben, *Homo Sacer*, p. 111, p. 115.

ly—the thanatopolitical expressions of sovereignty (epitomized by the Nazis' extermination of the Jews for a mere 'capacity to be killed' inherent in their condition as such)[16] and finally unmasks its hidden logic.

On this basis, what is scarcely investigated, or altogether overlooked, by countless analyses of the notion of *homo sacer* is the very fact that, beginning with the introduction of the first volume of his series, Agamben explicitly relates such notion to the possibility of a 'new politics'.[17] Conversely, a new politics is unthinkable without an in-depth engagement with the historico-ontological dimension of *sacratio* and the structural political ambiguity of the state of exception. Although such new politics 'remains largely to be invented', very early on in *Homo Sacer*, Agamben unhesitatingly defines it as 'a politics no longer founded on the *exceptio* of bare life'.[18] Beyond the exceptionalist logic—by now self-imploded—that unites sovereignty to bare life, Agamben seems to envisage a *relational* politics that would succeed in 'constructing the link between *zoe* and *bios*'.[19] This link between the bare life of man and his political existence would 'heal' the original 'fracture' which is at the same time precisely what causes their progressive indistinction in the generalized state of exception. Having said this, Agamben also conceives of such new politics as a *non-relational* relation that 'will [...] have to put the very form of relation into question, and to ask if the political fact is not perhaps thinkable beyond relation and, thus, no longer in the form of a connection'.[20]

While here Agamben runs the risk of blatantly contradicting himself—at least terminologically—what appears to emerge from both these formulations is the cautious delineation of a *positive* figure of contemporary *homo sacer* (whom we all virtually are). This would be the one who would, if not overcome, then subvert ('put into question') the *form* in which the relation between bare life and political existence has so far been both thought and lived in the West. Even a rapid account of some of the different embodiments of *homo sacer* which Agamben takes as paradig-

---

16. Agamben, *Homo Sacer*, p. 114.
17. Agamben, *Homo Sacer*, p. 11.
18. Agamben, *Homo Sacer*, p. 11.
19. Agamben, *Homo Sacer*, p. 11.
20. Agamben, *Homo Sacer*, p. 29.

matic of twentieth-century biopolitics and its state of generalized exception no doubt allows us to give them opposite signs. On the one hand, the 'overcomatose person' negatively represents 'a purely bare life, entirely controlled by man and his technology' for which there is 'a stage of life beyond the cessation of all vital functions'.[21] On the other hand, the 'figure of life' of the Rwandan refugee, similarly defined as 'a figure of bare or sacred life',[22] positively preludes 'a politics in which bare life is no longer separated and excepted', and the notion of nationality is constructively transformed into 'the being-in-exodus of the citizen'.[23] Again, we must conclude that the very same historico-ontological unfolding of biopolitics, the coming to light of the (repressed) sovereign relation that is both a political and metaphysical destiny, produces radically divergent results. At this point, we could possibly agree with Alain Badiou who, in his recent *Logiques des mondes*, benevolently criticizes Agamben for conceiving 'being as weakness', a weakness which, at the same time, corresponds to 'the delicate, almost secret, persistence of life, that which remains to one who has nothing left'.[24] Badiou completely misunderstands the notion of bare life when, in stark opposition to Agamben's formula, he defines it as 'always sacrificed'.[25] However, he is probably right in suggesting that *homo sacer*, 'the one who is led back to his pure being *qua* transitory living being [*vivant transitoire*]', is ultimately, for Agamben, nothing less than the *'hero'* of politics.[26]

Most importantly, in order to capture the internal movements and possible contradictions of the political hierarchy of *sacratio* implicitly proposed by Agamben, we should pay particular attention to the figure of the *Muselmann*, 'the most extreme figure' of the Nazi concentration camp inhabitant. Precisely because he has lost all consciousness and all personality and lives in an 'absolutely apathetic' way due to the humiliation, horror, and fear he has suffered, the *Muselmann* also surprisingly embodies 'a silent form of resistance'.[27] Even more problematically, Agamben

---

21. Agamben, *Homo Sacer*, p. 164, p. 161.
22. Agamben, *Homo Sacer*, p. 133.
23. Agamben, *Homo Sacer*, p. 134. Also Agamben, *Means Without End*, p. 25.
24. Alain Badiou, *Logiques des mondes: L'être et l'événement 2*, Paris, Seuil, 2006, p. 583.
25. Badiou, *Logiques des mondes*, p. 584.
26. Badiou, *Logiques des mondes*, p. 584 (my emphasis).
27. Agamben, *Homo Sacer*, p. 185.

seems to propose the *Muselmann* as a *paradigmatic* form of resistance to the logic of sovereign exception. A close reading of the last six pages of *Homo Sacer* allows us to neatly distinguish the twentieth-century *Muselmann* from the *homo sacer* of Roman law: while the latter is irremediably 'caught' by the very same power that bans him, the former manages to 'threaten' the law of the camp.[28] While the Roman *homo sacer*, in being pure *zoe*, pure bare life, founds the biopolitics of sovereign exception, the *Muselmann*, in *not* being pure *zoe* but rather 'an absolute indistinction of fact and law, of life and juridical rule, and of nature and politics', renders biopolitics literally power-less.[29] ('The guard suddenly seems *powerless* before [the *Muselmann*]', Agamben says.)[30] We are thus left to conclude that not only should biopolitics be understood as a necessary historico-ontological destiny but that we can prepare the overcoming of its exceptionalist logic of sovereignty exclusively within the horizon of the generalized biopolitics of Nazism—culminated in the extermination of Jews.

Agamben further outlines this ambiguously positive political dimension of *homo sacer* by means of two other notions: Heidegger's facticity and Benjamin's messianism. Their unexpected overlapping seems to provide us with a theoretical tool that both throws light on and complicates the concrete figure of the *Muselmann*. Already in *Homo sacer*, Agamben straightforwardly defines messianism as a 'theory of the state of exception'.[31] More precisely, following Benjamin, the messianic man is the one who brings about, 'realizes', a state of exception that has as yet remained only 'ideological', or 'virtual'. Acknowledging that the state of exception has turned into a rule, and the law is being in force without significance, the messianic man opposes such 'form of law' that continues to let 'bare life subsist before it' to a *form of life* for which politics is no longer thought in the form of a relation.[32] In other words:

> Law that becomes indistinguishable from life in a *real* state of exception is confronted by life that, in a symmetrical but *inverse* gesture, is entirely transformed into law. [...] Only at

---

28. Agamben, *Homo Sacer*, pp. 183-5.
29. Agamben, *Homo Sacer*, p. 185.
30. Agamben, *Homo Sacer*, p. 185 (my emphasis).
31. Agamben, *Homo Sacer*, pp. 57-8.
32. Agamben, *Homo Sacer*, pp. 53-5. See also p. 60.

this point do the two terms distinguished and kept united by the [sovereign] relation [...] (bare life and the form of law) abolish each other and enter into a new dimension.[33]

Agamben believes that he can recover a similar concept of form of life, a non-relational relation by means of which the sovereign relation based on the inclusive exclusion of bare life is overcome, in Heidegger's notion of factical life. Like Benjamin's notion of messianism, Heidegger's *faktisches Leben* would anticipate and pave the way to a new non-relational politics. And yet, this cannot occur without Heidegger developing a notion of life that is initially alarmingly proximate to that of National Socialism:

> For both Heidegger and National Socialism, life has no need to assume 'values' external to it in order to become politics: life is immediately political in its very facticity. [...] Man is not a duality of spirit and body, nature and politics, life and *logos*, but is instead resolutely situated at the point of their indistinction.[34]

Having said this, while Nazism eugenically resolves facticity into the 'incessant decision' on what is *sacer*—'life that does not deserve to live'—Heidegger makes it correspond to a suspension of all decisions concerning life, that is, an acknowledgement of the impossibility of isolating bare life. From this perspective, *Dasein* is nothing else than a form of life in which there is an 'inseparable unity of Being and ways of being', a positive *homo sacer* whose essence entirely lies [*liegt*] in existence and over whom, consequently, 'power no longer seems to have any hold'.[35]

---

33. Agamben, *Homo Sacer*, p. 55 (my emphases). See also Agamben, *Means Without End*, pp. 3-12.

34. Agamben, *Homo Sacer*, p. 153. On this point, see also Agamben's introduction to Emmanuel Lévinas, *Alcune riflessioni sulla filosofia dell'hitlerismo*, Macerata, Quodlibet, 1997.

35. Agamben, *Homo Sacer*, p. 153. See also p. 188. Interestingly enough, in *The Time That Remains*, Agamben criticizes in passing Heidegger's notion of facticity insofar as it still underlies the idea of appropriation. For Heidegger, the decisive element of human existence as factical life is an appropriation of the improper. According to Agamben, Heidegger therefore fails to understand facticity in terms of *use*. This is precisely what differentiates Heideggerian facticity from a messianic form of life which Agamben now conceives of in Christian—Pauline—terms (Giorgio Agamben, *The Time That Remains: A Commentary on the Letter to the Romans*, trans. Patricia Dailey, Stanford, Stanford University Press, 2005, p. 34).

But, leaving aside the theoretical formulations of philosophy, how and when is this passage from negative to positive *homo sacer* historically accomplished for Agamben? If we return to the 'figure of life' of the *Muselmann*, we soon realize that Agamben leaves this fundamental question unanswered in *Homo Sacer*, or, more correctly, that he provides us with contradictory answers to it. We have already seen how the *Muselmann* may be regarded as a paradigm of anti-biopolitical resistance: in this sense, he is a form of life that actively opposes the sovereign relation and, as shown by his alleged influence over the camp guard, even manages to neutralize its power. Surprisingly enough, Agamben does not problematize here the fact that the *Muselmann's* being an emancipatory form of life practically equates with his inability of 'distinguishing between pangs of cold and the ferocity of the SS'.[36] It is precisely the identification of cold with the SS, his 'mov[ing] in an absolute indistinction of fact and law', that makes him resistant to Nazism. Yet, at the same time, Agamben readily observes that, given the *Muselmann's* absolutely apathetic condition, 'nothing "natural" [...] remains in his life'.[37] In this sense, not only should the *Muselmann* be positively distinguished from the pure *zoe* of the Roman *homo sacer* who remains caught in power, but, having quite simply lost his instincts, he must also be negatively separated from any emancipatory form of life. As Agamben clearly states in the very last page of *Homo Sacer*, the form of life is indeed to be conceived of as a *bios* that is only its own *zoe*, a 'life that, being its own form, remains inseparable from it'.[38] In other words, no emancipatory form of life can be reduced to what cancels instinctual life. We are thus left with an impasse concerning the political value of the *Muselmann*. In short: is the *Muselmann qua* passage between negative and positive *sacratio* on the side of the overcomatose *homo sacer* (but then, why would he be a 'resistant'?) or on that of the refugee (but then, how to account for the gap that separates him from the form of life)? Is such obligatory transit after all thinkable? These are the two basic questions Agamben both unintentionally formulates and leaves undecided in *Homo Sacer*.

---

36. Agamben, *Homo Sacer*, p. 185.
37. Agamben, *Homo Sacer*, p. 185.
38. Agamben, *Homo Sacer*, p. 188.

2. *HOMO SACER*: A FRANCISCAN ONTOLOGY

Five years after *Homo Sacer*, Agamben further elaborated his investigation of biopolitics in the book he dedicated to Saint Paul, *The Time That Remains* (2000). We could suggest that, in this volume, the figure of *homo sacer* as earthly hero is transposed onto that of the messianic *Christian* man:[39] such idiosyncratic development of the 'Muslim' Jew analysed in *Homo Sacer* should be conceived beyond both Benjamin's non-Christian messianism and a merely analogical use of the messianic. On the one hand, Agamben carries out an unexpected Christianisation of Benjamin showing how his Second Thesis on history was supposedly derived from Paul's Second Letter to the Corinthians.[40] On the other hand, unlike other thinkers who recently appropriated the Apostle's works as a metaphoric example of political militancy,[41] Agamben believes that today's generalized state of exception should *directly* be understood in messianic terms. To cut a long story short, Christian messianic time is to be considered as the 'paradigm' of historical time, 'the only real time'.[42]

Criticising a common misunderstanding, Agamben asserts that messianic time should not be identified with the apocalyptic end of time. Messianic time is rather '*the time of the end* [...] the time that contracts itself and begins to end [...] the time that remains between time and its end'.[43] Developing a relation he had already introduced in *Homo Sacer* with regard to Benjamin, Agamben seems to suggest that such Christian messianic time could also be understood as a time of passage between the generalisation of the state of exception and its overcoming in a new non-relational politics. Like the state

---

39. In opposition to the Church's obliteration of Paul's alleged messianism, Agamben maintains that Christianity—the religion of the followers of *Christos*, literally, 'the Messiah'—can only be messianic (Agamben, *The Time That Remains*, pp. 15-6).

40. Agamben, *The Time That Remains*, pp. 138-45.

41. See Alain Badiou, *Saint Paul: The Foundation of Universalism*, trans. Ray Brassier, Stanford, Stanford University Press, 2003, Slavoj Žižek, *The Fragile Absolute—or, Why is the Christian legacy worth fighting for?*, London, Verso, 2000, Slavoj Žižek, *The Puppet and the Dwarf: The Perverse Core of Christianity*, Cambridge, MIT, 2003.

42. Agamben, *The Time That Remains*, p. 3, p. 6.

43. Agamben, *The Time That Remains*, p. 62.

of exception, Pauline Messianism suspends the law from within the law and consequently fulfils it. Instead of simply negating the rules of the existing order, Messianic law as the law of faith deactivates them in the form of the 'as not' [*hos me*]. As Agamben writes, 'to be messianic, to live in the Messiah, signifies the depropriation [*depropriazione*] of each and every juridical-factitious property (circumcized/uncircumcized; free/slave; man/woman) under the form of the *as not*'.[44] A Christian should live as if he were not that which he is according to the existing order—say, a free, uncircumcized man—whilst remaining within that very order. Such depropriation of law does not amount to a new identity. Messianic life is rather a use [*kresis*]: 'The messianic vocation is not a right, nor does it constitute an identity: it is a generic power [*potenza*] that one can use without ever being its proprietor'.[45]

Beyond *Homo Sacer*, the Pauline framework of *The Time That Remains* enables Agamben to think exhaustively the temporal complexity that characterizes messianism. Messianic time as the time that it takes for time to finish is not simply a segment added to the line of chronological time.[46] It is not sufficient to think of it as the time in between Christ's resurrection and his final coming at the end of time, the *parousia* that coincides with the Apocalypse. Messianic time should rather be equated with the time we need to 'bring to an end, to achieve our representation of time'.[47] From this perspective, eschatological and chronological time can no longer be clearly distinguished: the *kairos* 'is nothing else than a *chronos* that is grasped' as such.[48] In theological terms, this can only imply that the *para-ousia*—which after all means 'presence', literally a being that in Heideggerian fashion 'lies' by itself—does not correspond to the 'second coming' of Jesus. The Messiah is here; his return already contained in the event of resurrection: Christians only need a remainder of time to acknowledge the fact that the inoperativity of earthly laws is already operational.[49] From this also follows that Christians

---

44. Agamben, *The Time That Remains*, p. 26 (my translation).
45. Agamben, *The Time That Remains*, p. 26 (my translation).
46. See Agamben, *The Time That Remains*, p. 67.
47. Agamben, *The Time That Remains*, p. 67.
48. Agamben, *The Time That Remains*, p. 69 (my translation).
49. Agamben, *The Time That Remains*, pp. 70-1.

should fight against any authority that 'contrasts and conceals' such messianic state of anomie. More precisely, the *katechon*, the constituted power that defers the revelation of the messianic inoperativity of earthly laws, is the very same power that will retroactively appear as the supreme *anomos*, the Anti-Christ, once the Messiah's *parousia* will fully be assumed.[50]

While Agamben's arguments clearly invite us to map Paul's superimposition of the *katechon* over the *anomos* back onto Nazism (whose juridical status is indeed unsurprisingly defined in *State of Exception* as that of a 'legal civil war'),[51] this short-circuit raises a number of new important questions. Do we still inhabit the very same radical state of exception inaugurated by Christ's resurrection? How can we then account for what Agamben deems to be its increasing generalisation in modernity and post-modernity?[52] Quite bluntly, does Agamben believe that the full extent of the *anomos* of the *katechon* was finally disclosed at Auschwitz? Given that such full disclosure—which, for Agamben, is the 'being-in-act of Satan in any power [*potenza*]'—[53] can only be brought about by a concomitant final Christian *parousia* (*qua* assumption of the messianic inoperativity of earthly laws), should we bitterly conclude that Auschwitz is the *Christian* event *par excellence*? On the other hand, if, more plausibly, the *anomos* of the *katechon* has not as yet been brought completely to light, must its complete revelation *necessarily* coincide with a biopolitical 'catastrophe' of vaster proportion than the extermination of Jews?[54] Most importantly, can we really not avoid such disaster? Wouldn't the elaboration of a *post*-Christian interpretation of the notion of Messianic *parousia* represent the minimal precondition for defusing the Apocalypse? And, similarly, why should we relate

---

50. Agamben, *The Time That Remains*, p. 111.

51. Giorgio Agamben, *State of Exception*, trans. Kevin Attell, Chicago, University of Chicago Press, 2005, p. 2. See also Agamben, *The Time That Remains*, pp. 105-6.

52. Agamben claims that Paul himself already '*radicalizes* the condition of the state of exception' (Agamben, *The Time That Remains*, p. 106 [my emphasis]). Should we then infer that the contemporary radicalisation of the state of exception is the radicalisation of a radicalisation that has been ongoing for two thousand years?

53. Agamben, *The Time That Remains*, p. 111 (my translation).

54. This is what Agamben seems to suggest in *State of Exception*, pp. 86-87.

what is at stake in all the above questions to the *historico-ontological* unfolding of the inclusively exclusive capture of bare life carried out by sovereign power? Shouldn't we rather attempt to think the connection between being and the sovereign relation—the being-involved in the sovereign relation—differently, that is, beyond Heidegger?

Leaving aside these further complications of Christian temporality—which seem to both solve the paradoxes of *homo sacer* as Jewish *Muselmann* and make them re-emerge at a different level—we should finally focus on the issue of the compatibility between Paul's 'time that remains' and positive biopolitics. It must be said that, in the volume he dedicates to the Apostle, Agamben does not ever explicitly refer to biopolitics or *homo sacer*. In addition to this, he strangely fails to comment on the well-known passages of the Letter to the Romans in which Paul analyses the way in which life and death interact with the advent of the law (7:7-13).[55] Nevertheless, it is doubtless the case that a positively biopolitical dimension underlies Agamben's Pauline messianism insofar as, for him, the messianic manages to reverse the sovereign nexus between power and life. Messianism ultimately resolves itself into the non-relational relation of a 'form of life'.[56] In *The Time That Remains*, such notion is clearly ascribed a number of Christian theological attributes that it did not possess in *Homo Sacer* (and *Means without End*):

1. Messianic life as form of life should be understood in terms of *grace* [*charis*], that is, 'the capacity to [...] carry out good works independently of the law'.[57] In messianism there cannot be any conflict between different powers: grace as form of life emerges from a 'disconnection' [*sconnessione*] of (the opposition between) existing powers that interrupts current 'exchange and social obligations'.[58] Such disconnecting interruption represents as such a new kind of sovereignty [*autarkeia*] diametrically opposed to the sovereignty exercised by the anomic form of law.[59]

---

55. These passages are analysed by Badiou in *Saint Paul*, pp. 82-83. For Žižek's critique of Badiou's reading, see Slavoj Žižek, *The Ticklish Subject: The Absent Centre of Political Ontology*, London, Verso, 1999, pp. 145-51.
56. See Agamben, *The Time That Remains*, p. 122.
57. Agamben, *The Time That Remains*, p. 121.
58. Agamben, *The Time That Remains*, p. 120, p. 124.
59. See Agamben, *The Time That Remains*, pp. 120-1.

2. Grace as form of life must be strictly related to *faith* [*pistis*] as 'an experience of the word'.[60] Professing one's faith is a self-referential speech whose effectiveness relies on its being performed. Beyond performativity, the word of faith ultimately amounts to a 'pure and common *power* of saying [*potenza di dire*]' that both refuses any 'content of faith'—and could thus be regarded as 'weak'— and does not exhaust itself in 'the act of saying'.[61] 'From the perspective of faith, to hear a word does not entail ascertaining the truth of a given semantic content, nor does it simply entail renouncing understanding. [...] Neither a glossolalia deprived of meaning, nor mere denotative word, the word of faith enacts its *meaning* through its being *uttered*'.[62]
3. The word of faith as form of life corresponds to the *law of faith* [*kaine diatheke*], the new law of the Gospel that renders inoperative both Roman and Mosaic laws. Such law is, first and foremost, not a written text but 'the very life of the Messianic community'.[63] Any reduction of the Gospel to a form of law, a set of normative precepts, should be regarded as a betrayal of faith in the Messiah.

On the basis of such a detailed Christian development of the notion of 'form of life', I find it difficult to agree with Roberto Esposito's persistent attempt to confine Agamben's thought to the field of a negative critique of biopolitics. Esposito's elaboration of a philosophy that would depart from both Agamben's reduction of biopolitics to 'an antinomic repetition of the sovereign power's lethal paradigm' and Negri's identification of biopolitics with 'a power of life that is always excessive and finally subversive' is commendable and to a large degree successful.[64] However, I be-

---

60. Agamben, *The Time That Remains*, p. 129.
61. Agamben, *The Time That Remains*, pp. 136-137 (my translation; my emphasis)
62. Agamben, *The Time That Remains*, p. 129, p. 131 (my translation; my emphases).
63. Agamben, *The Time That Remains*, p. 122.
64. Roberto Esposito, 'Prefazione', in Laura Bazzicalupo, *Il governo delle vite: Biopolitica ed economia*, Bari-Roma, Laterza, 2006, p. VII. For a concise and exceptionally clear introduction to Esposito's work, see Roberto Esposito, 'Biopolitica, immunità, comunità', in A. Cutro (ed.), *Biopolitica: Storia e attualità di un concetto*, Verona, Ombre Corte, 2005, pp. 158-67. For Esposito's most refined

lieve that Esposito should pay more attention to the far from coincidental fact that his own deliberately 'affirmative biopolitics'[65] culminates in a notion, that of 'norm of life', which is undeniably contiguous to Agamben's notion of 'form of life'. Recovering a messianic dimension in Paul's Letters, Agamben is certainly able to configure a positive biopolitics: as we have just observed, *charis*, *pistis*, and *kaine diatheke* allow us to delineate new concepts of sovereignty, power, and law which, as forms of life, definitely undo the inclusive exclusion of bare life. The possibility that Agamben originally intended to keep messianism and biopolitics apart—in this sense, messianism would always-already represent an overcoming of biopolitics and the latter would, by definition, be negative—becomes at this stage irrelevant and, after all, profoundly incompatible with his detailed analysis of non-linear Christian temporality. Briefly, if following Agamben's own arguments, the generalisation of the state of exception—in which we have possibly lived since Christ's resurrection—is already retroactively messianic, then there *must* be a positive biopolitics.

Having said that, the fact remains that Agamben is able to formulate a transvaluation of biopolitics only in the guise of a bio-theo-politics. The importance of this conclusion cannot be overstated. Badiou is therefore correct in emphasising that Agamben's thought ultimately expresses a 'latent Christianity' for which the heroic *homo sacer* of politics is silently turned into the *homo messianicus* of Christian religion. Furthermore, according to this interpretation, Agamben's notion of 'weak' [*faible*] being, a being characterized by a 'presentative poverty', could qualify his ontology as 'Franciscan'. Although Badiou's remarks are concentrated in less than two pages, this appellation seems far from gratuitous, especially once we give the right weight to what Agamben himself says about Franciscanism in *The Time That Remains*. Francis and his followers conceive their Order as a 'messianic community', Agamben claims, whose ultimate aim is to 'create a space that escaped the grasp of power and its laws, without entering into conflict with them yet rendering them inoperative'.[66] This can be achieved by means of the so-

---

analysis of biopolitics to date, see Roberto Esposito, *Bíos: biopolitica e filosofia*, Torino, Einaudi, 2004.

65. Esposito, *Bíos*, p. XVI.
66. Agamben, *The Time That Remains*, p. 27.

called *usus pauper*, literally 'the poor use', which Agamben unhesitatingly defines, again, as a 'form of life'.[67] In other words, the Franciscan principle of poverty does not limit itself to refusing private property, but rather promotes a use of worldly goods that, as ontological 'nullification' (the 'as not'/'*hos me*'),[68] radically subtracts itself from the sphere of civil law. Here Agamben's distinction between 'imperfect nihilism' and 'messianic nihilism', which in *Homo Sacer* he derives from Benjamin, finds its final Christian meaning. Just like *homo sacer*—in the guise of the Jewish *Muselmann*—the Franciscan poor suspends the inclusively exclusive relation between sovereignty and bare life assuming their inextricability, but beyond him he also *vitally* inverts their sequence. The messianic Franciscan confronts the form of law of sovereign power with the form of life of Christ *qua* Gospel: as Agamben observes, *haec est vita evangeli Jesu Christi* ('This is the *life* of Jesus Christ's Gospel') is indeed the first rule of the Franciscan order.[69]

---

67. Agamben, *The Time That Remains*, p. 27.
68. Agamben, *The Time That Remains*, p. 23.
69. Agamben, *The Time That Remains*, p. 27. It is worth recalling that the very last paragraph of Negri and Hardt's *Empire* is dedicated to Saint Francis and the way in which his anti-instrumental adoption of poverty as 'ontological power' allegedly contributes to the emergence of a 'new society'. 'There is an ancient legend that might serve to illuminate the future life of communist militancy: that of Saint Francis of Assisi. Consider his work. To denounce the poverty of the moltitude he adopted that common condition and discovered there the ontological power of a new society. [...] Francis in opposition to nascent capitalism refused every instrumental discipline, and in opposition to the mortification of the flesh (in poverty and in the constituted order) he posed a joyous life [...]. Once again in postmodernity we find ourselves in Francis's situation, posing against the misery of power the joy of being' (Michael Hardt and Antonio Negri, *Empire*, Cambridge, Harvard University Press, 2000, p. 413). Instead of doxastically opposing Negri's positive biopolitics to Agamben's negative biopolitics, we should explore to what extent these authors' theories overlap and find common *Christian* references.

REFERENCES

Agamben, Giorgio, *Il linguaggio e la morte. Un seminario sul luogo della negatività*, Turin, Einaudi, 1982.
Agamben, Giorgio, *Language and Death: The Place of Negativity*, trans. Karen E. Pinkus with Michael Hardt, Minneapolis, University of Minnesota Press, 1991.
Agamben, Giorgio, *The Coming Community*, trans. Michael Hardt, Minneapolis, University of Minnesota Press, 1993.
Agamben, Giorgio, *Homo sacer. Il potere sovrano e la nuda vita*, Turin, Einaudi, 1995.
Agamben, Giorgio, *Homo Sacer: Sovereign Power and Bare Life*, trans. Daniel Heller-Roazen, Stanford, Stanford University Press, 1998.
Agamben, Giorgio, *Means Without End: Notes on Politics*, trans. Vincenzo Binetti and Cesare Casarino, Minneapolis, University of Minnesota Press, 2000.
Agamben, Giorgio, *The Time That Remains: A Commentary on the Letter to the Romans*, trans. Patricia Dailey, Stanford, Stanford University Press, 2005.
Agamben, Giorgio, *State of Exception*, trans. Kevin Attell, Chicago, University of Chicago Press, 2005.
Agamben, Giorgio, *Il Regno e la Gloria: Per una genealogia teologica dell'economia e del governo*, Milano, Neri Pozza, 2007.
Anderson, Perry, 'An Invertebrate Left', *London Review of Books*, vol. 31, no. 5, 12 March 2009.
Anderson, Perry, 'An Entire Order Converted into What it Was Intended to End', *London Review of Books*, vol. 31, no. 4, 26 February 2009.

Apel, Karl-Otto, *Towards a Transformation of Philosophy*, trans. D. Frisby and G. Adey, London, Keagan Paul International, 1980.
Arendt, Hannah, *On Violence*, New York, Harcourt & Brace, 1970.
Badiou, Alain, *Saint Paul: The Foundation of Universalism*, trans. Ray Brassier, Stanford, Stanford University Press, 2003.
Badiou, Alain, *Logiques des mondes: L'être et l'événement 2*, Paris, Seuil, 2006.
Bataille, Georges, *On Nietzsche*, trans. Bruce Boone, London, Athlone Press, 1992.
Blanchot, Maurice, *La communauté inavouable*, Paris, Minuit, 1984.
Bologna, Sergio, 'A Review of Storming Heaven: Class Composition and Struggle in Italian Autonomist Marxism by Steve Wright', *Strategies: Journal of Theory, Culture & Politics*, vol. 16, no. 2, 2003.
Borio, Guido, Francesca Pozzi and Gigi Roggero, *Gli operaisti*, Roma, DeriveApprodi, 2005.
Borradori, Giovanna (ed.), *The New Italian Philosophy*, Evanston, Northwestern University Press, 1989.
Cacciari, Massimo, 'Noi, i soggetti', *Rinascita*, no. 27, 2 July 1976.
Cacciari, Massimo, 'Sulla genesi del pensiero negativo', *Contropiano*, vol. 1, 1969.
Cacciari, Massimo, *Krisis. Saggio sulla crisi del pensiero negativo da Nietzsche a Wittgenstein*, Milan, Feltrinelli, 1976.
Cacciari, Massimo, '"Razionalità" e "Irrazionalità" nella critica del Politico in Deleuze e Foucault', *Aut Aut*, no. 161, 1977, pp. 119-33.
Cacciari, Massimo, 'Critica della "autonomia" e problema del politico', in V.F. Ghisi (ed.), *Crisi del sapere e nuova razionalità*, Bari, De Donato, 1978, pp. 123-35.
Cacciari, Massimo, *Dallo Steinhof. Prospettive viennesi del primo Novecento*, Milan, Adelphi, 1980.
Cacciari, Massimo, *Posthumous People: Vienna at the Turning Point*, trans. R. Friedman, Stanford, Stanford University Press, 1996.
Calarco, Matthew and Steven DeCaroli (eds.), *Sovereignty and Life: Essays on the Work of Giorgio Agamben*, Stanford, Stanford University Press, 2007.

Cantarano, Giuseppe, *Immagini del nulla. La filosofia italiana contemporanea*, Milan, Bruno Mondatori, 1998.
Cantarano, Giuseppe, *Immagini del nulla. La filosofia italiana contemporanea*, Milano, Bruno Mondadori, 1998.
Chomsky, Noam, *Language and the Problems of Knowledge: The Managua Lectures*, Cambridge, MIT Press, 1988.
Cleaver, Harry, *Reading Capital Politically*, 2nd ed., London, AK Press, 2000.
Corradi, Cristina, *Storia dei marxismi in Italia*, Roma, manifestolibri, 2005.
Davis, Creston, John Milbank and Slavoj Žižek (eds.), *Theology and the Political: The New Debate*, Durham, Duke University Press, 2005.
De Giorgi, Alessandro, *Zero Tolleranza. Strategie e pratiche della società di controllo*, Rome, DeriveApprodi, 2000.
De Giorgi, Alessandro, *Il governo dell'eccedenza. Postfordismo e controllo della moltitudine*, Verona, Ombre Corte, 2002.
De Martino, Ernesto, *La fine del mondo. Contributo all'analisi delle apocalissi culturali*, Torino, Einaudi, 2002 [1977].
Esposito, Roberto, *Communitas. Origine e destino della comunità*, Turin, Einaudi, 1998.
Esposito, Roberto, *Bíos: biopolitica e filosofia*, Torino, Einaudi, 2004.
Esposito, Roberto, 'Biopolitica, immunità, comunità', in A. Cutro (ed.), *Biopolitica: Storia e attualità di un concetto*, Verona, Ombre Corte, 2005, pp. 158-67.
Esposito, Roberto, 'Prefazione', in Laura Bazzicalupo, *Il governo delle vite: Biopolitica ed economia*, Bari-Roma, Laterza, 2006.
Fumagalli, Andrea, Christian Marazzi and Adelino Zanini, *La moneta nell'Impero*, Verona, Ombre Corte, 2002.
Gehlen, Arnold, *Philosophische Anthropologie und Handlungsleghre*, Frankfurt am Main, Klostermann, 1983.
Givone, Sergio, *Storia del nulla*, Bari, Laterza, 1995.
Gould, Stephen Jay, *Ontogeny and Phylogeny*, Cambridge, Belknap Harvard, 1977.
Hardt, Michael, 'Introduction: Laboratory Italy', in Paolo Virno and Michael Hardt (ed.), *Radical Thought in Italy: A Potential Politics*, Minneapolis, University of Minnesota Press, 1996.

Hardt, Michael and Antonio Negri, *Empire*, Cambridge, Harvard University Press, 2000.
Hardt, Paolo Virno and Michael (ed.), *Radical Thought in Italy: A Potential Politics*, Minneapolis, University of Minnesota Press, 1996.
Heidegger, Martin, *On Time and Being*, trans. Joan Stambaugh, Chicago, University of Chicago Press, 1972.
Heidegger, Martin, 'The Thing', in *Poetry, Language, Thought*, trans. Albert Hofstadter, London, Harper Perennial, 1976.
Heidegger, Martin, 'The Origin of the Work of Art', in David Farrell Krell (ed.), *Basic Writings*, London, Routledge and Kegan Paul, 1978.
Heidegger, Martin, 'Postscript to "What is Metaphysics?"', in *Pathmarks*, ed. and trans. William McNeill, Cambridge, Cambridge University Press, 1998.
Heidegger, Martin, 'What is Metaphysics?', in William McNeill (ed.), *Pathmarks*, trans. David F. Krell, Cambridge, Cambridge University Press, 1998.
Kelsen, Hans, *Vom Wesen und Wert der Demokratie*, Tübingen, Mohr, 1926.
Lévinas, Emmanuel, *Alcune riflessioni sulla filosofia dell'hitlerismo*, Macerata, Quodlibet, 1997.
Marx, Karl, *Early Writings*, trans. Rodney Livingstone and Gregor Benton, London, Penguin, 1975.
Marx, Karl, *Capital, Volume 1: A Critique of Political Economy*, trans. Ben Fowkes, London, Penguin, 1992.
Marx, Karl, *Grundrisse: Foundations of the Critique of Political Economy*, trans. Martin Nicolaus, London, Penguin, 1993.
Marx, Karl, *Il Capitale, Libro I, Capitolo VI Inedito*, Bruno Maffi (ed.), Milan, Etas, 2002.
Marx, Karl and Friedrich Engels, *The German Ideology*, Moscow, Progress Publishers, 1976.
Mazzeo, Marco, *Tatto e linguaggio*, Rome, Editori Riuniti, 2004.
Mezzadra, Sandro, *Diritto di fuga. Migrazioni, cittadinanza, globalizzazione*, 2nd ed., Verona, Ombre Corte, 2006.
Muraro, Luisa, 'Libreria delle Donne di Milano', *Via Dogana*, no. 86, 2008.
Muraro, Luisa, *Al mercato della felicità. La forza irrinunciabile del desiderio [At the Market of Happiness: The Unrenounceable Force of Desire]*, Milan, Mondadori, forthcoming 2009.

Nancy, Jean-Luc, *The Inoperative Community*, trans. Lisa Garbus Peter Connor, Michael Holland, and Simona Sawhney, Minneapolis, University of Minnesota Press, 1991.
Nancy, Jean-Luc, *The Sense of the World*, trans. Jeffrey S. Librett, Minneapolis, University of Minnesota Press, 1997.
Negri, Antonio, 'Oltre i confini della società del controllo', *il manifesto*, 15 June 2002.
Negri, Antonio, *Descartes politico. Della ragionevole ideologia*, Milan, Feltrinelli, 1970.
Negri, Antonio, *L'anomalia selvaggia*, Milan, Feltrinelli, 1981.
Negri, Antonio, 'Note sulla storia del politico in Tronti', in *L'anomalia selvaggia*, Milan, Feltrinelli, 1981.
Negri, Antonio, *La macchina tempo. Rompicapi Liberazione Costituzione*, Milan, Feltrinelli, 1982.
Negri, Antonio, *Fabbriche del soggetto*, Livorno, XXI Secolo, 1987.
Negri, Antonio, *Revolution Retrieved: Selected Writings on Marx, Keynes, Capitalist Crisis and New Social Subjects 1967–1983*, trans. Ed Emery and John Merrington, London, Red Notes, 1988.
Negri, Antonio, *Marx Beyond Marx: Lessons on the Grundrisse*, Jim Fleming (ed.), trans. Michael Ryan, Mauricio Viano, and Harry Cleaver, New York, Autonomedia, 1989.
Negri, Antonio, *The Savage Anomaly: The Power of Spinoza's Metaphysics and Politics*, trans. Michael Hardt, Minneapolis, University of Minnesota Press, 1991.
Negri, Antonio, *Lenta ginestra. Saggio su Leopardi*, Milano, Mimesis Eterotopia, 2001 [1987].
Negri, Antonio, 'Prefazione', in Francesco Lesce (ed.), *Un'ontologia materialista. Gilles Deleuze e il XXI secolo*, Milan, Mimesis, 2004.
Negri, Antonio, 'The Political Subject and Absolute Immanence', in Creston Davis, John Milbank, and Slavoj Žižek (eds.), *Theology and the Political: The New Debate*, trans. Matteo Mandarini, Durham, Duke University Press, 2005.
Negri, Antonio, 'Domination and Sabotage', in Timothy S. Murphy (ed.), *Books for Burning: Between Civil War and Democracy in 1970s Italy*, trans. E. Emery, New York, Verso, 2005.
Negri, Antonio, *Books for Burning*, Timothy S. Murphy (ed.), London, Verso, 2005.

Negri, Antonio, 'Giorgio Agamben: the Discreet Taste of the Dialectic', in Matthew Calarco and Steven DeCaroli (eds.), *Sovereignty and Life: Essays on the Work of Giorgio Agamben*, trans. Matteo Mandarini, Stanford, Stanford University Press, 2006, pp. xiii, 282 p.

Negri, Antonio, *I libri del rogo*, Roma, DeriveApprodi, 2006.

Negri, Antonio, *The Political Descartes: Reason, Ideology, and the Bourgeois Project*, trans. Matteo Mandarini and Alberto Toscano, London, Verso, 2007.

Parmenides and A. H. Coxon, *The Fragments of Parmenides: A Critical Text with Introduction, and Translation, the Ancient Testimonia and a Commentary*, trans. A. H. Coxon, Assen, Van Gorcum, 1986.

Peirce, Charles Sanders, *Collected Papers*, Charles Hartshorne and Paul Weiss (ed.), vol. 2, Cambridge, Harvard University Press, 1933.

Portmann, Adolf, *Aufbruch der Lebensforschung*, Zurich, Rhein Verlag, 1965.

Schmidt, Dennis J., *The Ubiquity of the Finite: Hegel, Heidegger, and the Entitlements of Philosophy*, Cambridge, MIT Press, 1988.

Schmitt, Carl, ed. and trans. Jeffrey Seitzer, Durham, Duke University Press, Constitutional Theory.

Schroeder, Silvia Benso and Brian (ed.), *Contemporary Italian Philosophy: Crossing the Borders of Ethics, Politics and Religion*, Albany, State University of New York, 2007.

Schürmann, Reiner, *Heidegger on Being and Acting: From Principles to Anarchy*, trans. Christine-Marie Gros, Bloomington, Indiana University Press, 1987.

Toscano, Alberto, 'Communism as Separation', in Peter Hallward (ed.), *Think Again: Alain Badiou and the Future of Philosophy*, London, Continuum, 2004.

Toscano, Alberto, 'Always Already Only Now', in T.S. Murphy and A.-K. Mustapha (eds.), *The Philosophy of Antonio Negri, Volume 2: Revolution in Theory*, London, Pluto, 2007.

Tronti, Mario, *Sull'autonomia del politico*, Milano, Feltrinelli, 1977.

Tronti, Mario, *Il tempo della politica*, Roma, Editori Riuniti, 1980.

Tronti, Mario, 'The Strategy of Refusal', in S. Lotringer and C.

Marazzi (eds.), *Autonomia: Post-Political Politics*, New York, Semiotext(e), 1980.
Tronti, Mario, *Con le spalle al futuro. Per un altro dizionario politico*, Roma, Editori Riuniti, 1992.
Tronti, Mario, *La politica al tramonto*, Torino, Einaudi, 1998.
Tronti, Mario, *Cenni di castella*, Fiesole, Cadmo, 2001.
Tronti, Mario, *Operai e capitale*, 3rd ed., Roma, DeriveApprodi, 2006.
Tse-Tung, Mao, *On the Correct Handling of Contradictions Among the People*, Peking, Foreign Languages Press, 1957.
Vattimo, Gianni, *The End of Modernity: Nihilism and Hermeneutics in Post-modern Culture*, trans. John R. Snyder, Cambridge, Polity Press, 1991.
Vattimo, Gianni, *Belief*, trans. Luca D'Isanto and David Webb, Cambridge, Polity Press, 1999.
Vattimo, Gianni, *After Christianity*, trans. Luca D'Isanto, New York, Columbia University Press, 2002.
Vigorelli, Amedeo, 'Noi, i soggetti e il "politico". A proposito di Bisogni e teoria', *Aut Aut*, no. 155-156, 1976, pp. 196-203.
Virno, Paolo, 'Do You Remember Counter-Revolution', in Paolo Virno and Michael Hardt (ed.), *Radical Thought in Italy: A Potential Politics*, Minneapolis, University of Minnesota Press, 1996.
Virno, Paolo, *Quando il verbo si fa carne. Linguaggio e natura umana*, Turin, Bollati Boringhieri, 2003.
Virno, Paolo, *Esercizi di esodo. Linguaggio e azione politica*, Verona, Ombre Corte, 2007.
Wittgenstein, Ludwig, *Tractatus Logico-Philosophicus*, trans. D. F. Pears and B. F. McGuinness, London, Routledge & Kegan Paul, 1961.
Wright, Steve, 'There and Back Again: Mapping the Pathways Within Autonomist Marxism', paper delivered at the *Immaterial Labour, Multitudes and New Social Subjects: Class Composition in Cognitive Capitalism*, Cambridge, 29-30 April 2006, paper available at: http://www.geocities.com/immateriallabour/wrightpaper2006.html.
Wright, Steve, *Storming Heaven: Class Composition and Struggle in Italian Autonomist Marxism*, London, Pluto, 2002.
Žižek, Slavoj, *The Ticklish Subject: The Absent Centre of Political Ontology*, London, Verso, 1999.

Žižek, Slavoj, *The Fragile Absolute—or, Why is the Christian legacy worth fighting for?*, London, Verso, 2000.

Žižek, Slavoj, *The Puppet and the Dwarf: The Perverse Core of Christianity*, Cambridge, MIT, 2003.

www.ingramcontent.com/pod-product-compliance
Lightning Source LLC
Chambersburg PA
CBHW022104160426
43198CB00008B/340